EVERY STUDENT CAN LEARN

MATHEMATICS
Assessment & Intervention
in a PLC at Work®
SECOND EDITION

Sarah Schuhl

Timothy D. Kanold

Mona Toncheff

Bill Barnes

Jessica Kanold-McIntyre

Matthew R. Larson

Georgina Rivera

A Joint Publication With

Copyright © 2024 by Solution Tree Press

Materials appearing here are copyrighted. With one exception, all rights are reserved. Readers may reproduce only those pages marked "Reproducible." Otherwise, no part of this book may be reproduced or transmitted in any form or by any means (electronic, photocopying, recording, or otherwise) without prior written permission of the publisher.

555 North Morton Street
Bloomington, IN 47404
800.733.6786 (toll free) / 812.336.7700
FAX: 812.336.7790

email: info@SolutionTree.com
SolutionTree.com

Visit **go.SolutionTree.com/MathematicsatWork** to download the free reproducibles in this book.

Printed in the United States of America

LCCN: 2023012434

Solution Tree
Jeffrey C. Jones, CEO
Edmund M. Ackerman, President

Solution Tree Press
President and Publisher: Douglas M. Rife
Associate Publishers: Todd Brakke and Kendra Slayton
Editorial Director: Laurel Hecker
Art Director: Rian Anderson
Copy Chief: Jessi Finn
Senior Production Editor: Suzanne Kraszewski
Proofreader: Charlotte Jones
Text and Cover Designer: Abigail Bowen
Acquisitions Editor: Hilary Goff
Assistant Acquisitions Editor: Elijah Oates
Content Development Specialist: Amy Rubenstein
Associate Editor: Sarah Ludwig
Editorial Assistant: Anne Marie Watkins

Acknowledgments

The new and enhanced chapters for this second edition of *Mathematics Assessment and Intervention in a PLC at Work* are based on the deep professional development our author team and Mathematics at Work™ associates have led since that book's publication in 2018. We have been privileged to work across North America in schools with mathematics educators using protocols and resources from the first edition. Questions, insights, and feedback from readers and users of our work are revealed in the examples and revisions needed to strengthen a teacher team's formative use of common assessments for teacher and student learning.

A special thank-you to Jeff Jones, Douglas Rife, Todd Brakke, Rian Anderson, and the incredible Suzanne Kraszewski, along with the editorial team that understands our vision for Mathematics in a PLC at Work. Your continued support for this second edition effectively guides mathematics educators in their pursuit to ensure students learn through the formative use of common assessments and intervention actions.

We deeply appreciate the feedback given to us by our colleagues and reviewers. Thank you for your insights and ideas to strengthen the rubrics, tools, examples, and protocols in the book.

We are also grateful to our families, who supported us as we learned from one another and crafted this book with new examples, tools, and revised rubrics. We are fortunate to have the support of loved ones as we do the work we so passionately pursue.

Finally, we are most grateful to you, the reader, and the work you are doing to ensure students learn mathematics through the formative use of common assessments. Students today need your wisdom and collective practices to grow a positive mathematics identity and develop the ability to productively reason when solving mathematical tasks. Your collaborative work to create and analyze common unit assessments will lead not only to improved student assessment practices and learning, but to effective instructional practices as well. Meaningful student reflection and action with targeted interventions and support will ensure the learning of each and every one of your students and serve as a vital routine within our *Every Student Can Learn Mathematics* series.

The authors would like to thank the following reviewers:

Kim Bailey
Author and Educational Consultant

Jacqueline Booker
Assistant Principal and Mathematics Specialist
Newhall Elementary School
Newhall, California

Brian Buckhalter
Solution Tree Associate

Suyi Chuang
Assistant Principal
Tuscarora High School
Ashburn, Virginia

Jennifer Deinhart
K–8 Mathematics Specialist
Solution Tree Author and Associate
Leesburg, Virginia

Maria Everett
Solution Tree Associate

Erin Lehmann
Assistant Professor
University of South Dakota
Rapid City, SD

Jennifer Rose Novak
Director of Curriculum, Instruction, and Assessment
Howard County Public Schools
Ellicott City, Maryland

Gwendolyn Zimmermann
Author and Educational Consultant

Visit **go.SolutionTree.com/MathematicsatWork** to download the free reproducibles in this book.

Table of Contents

About the Authors vii

Preface xi

Introduction 1

    Equity and PLCs 1
    The Reflect, Refine, and Act Cycle 2
    Mathematics in a PLC at Work Framework 2
    About This Book 4

1   The Mathematics at Work™ Common Assessment Process 5

    Four Specific Purposes 5
    The Assessment Instrument Versus the Assessment Process 7
    FAST Feedback 9
    Evaluation of Mathematics Assessment Feedback Processes 10
    Reflection on Team Assessment Practices 13

2   Quality Common Mathematics Assessments 17

    Identification of and Emphasis on Essential Learning Standards 21
    Balance of Higher- and Lower-Level-Cognitive-Demand Mathematical Tasks 26
    Variety of Assessment-Task Formats 30
    Appropriate and Clear Scoring Agreements 32
    Clarity of Directions, Academic Language, Visual Presentation, and Logistics 41

3   Sample Common Mathematics Assessments and Calibration Routines 47

    Sample Common Mid-Unit Assessments 47
    Sample Common End-of-Unit Assessments 53
    Collaborative Scoring Agreements 66
    Quality of Mathematics Assessment Feedback 76

4   Teacher Actions in the Formative Assessment Process 81

    Evidence of Student Learning Protocols 82

5 Student Actions in the Formative Assessment Process 97
Student Action After the Common End-of-Unit Assessment 97
Student Action After Common Mid-Unit Assessments 108
Student Trackers 108
Self-Regulatory Feedback 117

6 Team Response to Student Learning Using Tier 2 Mathematics Intervention Criteria 121
Systematic and Required 126
Targeted by Essential
Learning Standard 128
Fluid and Flexible 130
Just in Time 131
Proven to Show Evidence of Student Learning 132

Summary 135

Appendix 137
Cognitive-Demand-Level Task Analysis Guide 137
Mathematics Assessment and Intervention in a PLC at Work Protocols and Tools *138*

References and Resources 143

Index 147

About the Authors

Sarah Schuhl is an educational coach and consultant specializing in mathematics, professional learning communities, common formative and summative assessments, priority school improvement, and response to intervention (RTI). She has worked in schools as a secondary mathematics teacher, high school instructional coach, and K–12 mathematics specialist.

Schuhl was instrumental in the creation of a professional learning community (PLC) in the Centennial School District in Oregon, helping teachers make large gains in student achievement. She earned the Centennial School District Triple C Award in 2012.

Schuhl grows learning in districts throughout the United States through large-group professional development and small-group coaching. Her work focuses on strengthening the teaching and learning of mathematics, having teachers learn from one another when working effectively as a collaborative team in a PLC, and striving to ensure the learning of each and every student through assessment practices and intervention. Her practical approach includes working with teachers and administrators to implement assessments for learning, analyze data, collectively respond to student learning, and map standards.

For Mathematics at Work, Schuhl coauthored *Engage in the Mathematical Practices: Strategies to Build Numeracy and Literacy With K–5 Learners* and (with Timothy D. Kanold) the *Every Student Can Learn Mathematics* series and the *Mathematics at Work™ Plan Book*, and was editor of the *Mathematics Unit Planning in a PLC at Work* series. Additionally, she contributed to the National Council of Supervisors of Mathematics (NCSM) publication *NCSM Essential Actions: Framework for Leadership in Mathematics Education*. On the subject of priority schools, Schuhl coauthored *School Improvement for All: A How-To Guide for Doing the Right Work* and *Acceleration for All: A How-To Guide for Overcoming Learning Gaps*. She contributed to *Charting the Course for Leaders: Lessons From Priority Schools in a PLC at Work* and *Charting the Course for Collaborative Teams: Lessons From Priority Schools in a PLC at Work*.

Previously, Schuhl served as a member and chair of the National Council of Teachers of Mathematics (NCTM) editorial panel for the journal *Mathematics Teacher* and secretary of NCSM. Her work with the Oregon Department of Education includes designing mathematics assessment items, test specifications and blueprints, and rubrics for achievement-level descriptors. She has also contributed as a writer to a middle school mathematics series and an elementary mathematics intervention program.

Schuhl earned a bachelor of science in mathematics from Eastern Oregon University and a master of science in mathematics education from Portland State University.

To learn more about Sarah Schuhl's work, follow her @SSchuhl on Twitter.

Timothy D. Kanold, PhD, is an award-winning educator and author. He formerly served as director of mathematics and science and as superintendent of Adlai E. Stevenson High School District 125, a Model Professional Learning Community (PLC) district in Lincolnshire, Illinois.

Dr. Kanold has authored or coauthored and led more than thirty-seven textbooks and professional development books on K–12 mathematics, school culture, and school leadership, including his best-selling and 2018 Independent Publisher Book Award–winning book *HEART! Fully Forming Your Professional Life as a Teacher and Leader*. In 2021, he authored a sequel to *HEART!*, the book *SOUL! Fulfilling the Promise of Your Professional Life as a Teacher and Leader*, and most recently, he coauthored the best-selling book *Educator Wellness: A Guide for Sustaining Physical, Mental, Emotional, and Social Well-Being* (2022).

Dr. Kanold received the 2017 Ross Taylor / Glenn Gilbert National Leadership Award from the National Council of Supervisors of Mathematics, the international 2010 Damen Award for outstanding contributions to education from Loyola University Chicago, and the 1986 Presidential Award for Excellence in Mathematics and Science Teaching.

Dr. Kanold is committed to equity, excellence, and social justice reform for the improved learning of students and school faculty, staff, and administrators. He conducts inspirational professional development seminars worldwide with a focus on improving student learning outcomes in mathematics through a commitment to the PLC process. He also brings an intentional focus to how to live a well-balanced, fully engaged professional life by practicing reflection and self-care routines daily.

Dr. Kanold earned a bachelor's degree in education and a master's degree in applied mathematics from Illinois State University. He received his doctorate in educational leadership and counseling psychology from Loyola University Chicago.

To learn more about Dr. Kanold's work, follow him @tkanold, #heartandsoul4ED, #MAWPLC, or #liveyourbestlife on Twitter.

Mona Toncheff, an educational consultant and author, has over thirty years of experience in public education. She worked as both a mathematics teacher and a mathematics specialist for the Phoenix Union High School District in Arizona. In the latter role, she coached and provided professional development to high school teachers and administrators related to quality mathematics teaching and learning and effective collaborative teams. She currently serves as a supervisor teacher for the University of Arizona Teach Program.

Toncheff has supervised the culture change from teacher isolation to professional learning communities, created articulated standards and relevant district common assessments, and provided ongoing professional development on best practices, equity and access, technology, response to intervention, high-quality grading practices, and assessment for learning.

Toncheff has coauthored several Solution Tree and National Council of Supervisors of Mathematics (NCSM) publications. Her Solution Tree books include *Common Core Mathematics in a PLC at Work, High School* (2012); *Beyond the Common Core: A Handbook for Mathematics in a PLC at Work, High School* (2014); and *Activating the Vision: The Four Keys of Mathematics Leadership* (2016), and she is one of the lead coauthors of the *Every Student Can Learn Mathematics* series. Toncheff was the lead editor and contributing author for the NCSM books *Framework for Leadership in Mathematics Education* (2020), *Instructional Leadership in Mathematics Education* (2019), and *Culturally Relevant Leadership in Mathematics Education* (2022).

As a writer and consultant, Toncheff works with educators and leaders across the United States to build collaborative teams, empowering them with effective strategies for aligning curriculum, instruction, and assessment to ensure all students receive high-quality mathematics instruction.

A past president of NCSM, Toncheff has served as an active board member of NCSM for over eleven years. In addition, she was a cofounding board member and president of Arizona Mathematics Leaders. In 2009, she was named Phoenix Union High School District Teacher of the Year. In 2014, she received the Copper

Apple Award for leadership in mathematics from the Arizona Association of Teachers of Mathematics, and in 2022, she received the AML Leadership Award from Arizona Mathematics Leaders.

Toncheff earned a bachelor of science from Arizona State University and a master of education in educational leadership from Northern Arizona University.

To learn more about Mona Toncheff's work, follow her @toncheff5 on Twitter.

Bill Barnes is the chief academic officer for the Howard County Public School System in Maryland. He has held several positions on the National Council of Supervisors of Mathematics (NCSM) board of directors and has served as an adjunct professor for Johns Hopkins University, the University of Maryland–Baltimore County, McDaniel College, and Towson University.

Barnes is passionate about ensuring equity and access in mathematics for students, families, and staff. His experiences drive his advocacy efforts as he works to ensure opportunity and access for underserved and underperforming populations. He fosters partnerships among schools, families, and community resources in an effort to eliminate traditional educational barriers.

A past president of the Maryland Council of Teachers of Mathematics, Barnes has served as the affiliate service committee Eastern Region 2 representative for the National Council of Teachers of Mathematics (NCTM) and regional team leader for NCSM.

Barnes was the recipient of the 2003 Maryland Presidential Award for Excellence in Mathematics and Science Teaching. He was named Outstanding Middle School Math Teacher by the Maryland Council of Teachers of Mathematics and Maryland Public Television and Master Teacher of the Year by the National Teacher Training Institute.

Barnes earned a bachelor of science in mathematics from Towson University and a master of science in mathematics and science education from Johns Hopkins University.

To learn more about Bill Barnes's work, follow him @BillJBarnes on Twitter.

Matthew R. Larson, PhD, is an award-winning educator and author who served as the K–12 mathematics curriculum specialist for Lincoln Public Schools in Nebraska for more than twenty years. He served as president of the National Council of Teachers of Mathematics (NCTM) from 2016 to 2018. Dr. Larson has taught mathematics at the elementary through college levels and has held an honorary appointment as a visiting associate professor of mathematics education at Teachers College, Columbia University. He currently serves as the associate superintendent for instruction for Lincoln Public Schools, Nebraska.

Dr. Larson is coauthor of several mathematics textbooks, professional books, and articles on mathematics education, and was a contributing writer on the influential publications *Principles to Actions: Ensuring Mathematical Success for All* (NCTM, 2014) and *Catalyzing Change in High School Mathematics: Initiating Critical Conversations* (NCTM, 2018). A frequent keynote speaker at national meetings, Dr. Larson is well known for his humorous presentations and their application of research findings to practice.

Dr. Larson earned a bachelor's degree and doctorate from the University of Nebraska–Lincoln.

Jessica Kanold-McIntyre is an educational consultant and author committed to supporting teacher implementation of rigorous mathematics curriculum and assessment practices blended with research-informed instructional practices. She works with teachers and schools around the United States to meet the needs of their students. Specifically, she specializes in building and supporting the collaborative teacher culture through the curriculum, assessment, and instruction cycle.

Kanold-McIntyre is currently the executive director of human relations and superintendent-elect in Aptakisic-Tripp Community Consolidated School District 102, in Buffalo Grove, Illinois, where she has also served as a middle school principal, assistant principal, and mathematics teacher and leader. As principal of Aptakisic Junior High School, she supported her teachers in implementing initiatives, such as the Illinois Learning Standards; the Next Generation Science Standards; and the College, Career, and Civic Life Framework for Social Studies State Standards, while also supporting a one-to-one iPad environment for all students. She focused on teacher instruction through the PLC process, creation of learning opportunities around formative assessment practices, data analysis, and student engagement. She previously served as assistant principal at Aptakisic, where she led and supported special education, response to intervention (RTI), and English learner staff through the PLC process.

As a mathematics teacher and leader, Kanold-McIntyre strove to create equitable and rigorous learning opportunities for all students while also providing them engaging and challenging experiences to foster critical-thinking and problem-solving skills. As a mathematics leader, she developed and implemented a districtwide process for the Common Core State Standards in Illinois and led a collaborative process to create mathematics curriculum guides for K–8 mathematics, algebra 1, and algebra 2. She has served as a board member for the National Council of Supervisors of Mathematics (NCSM).

Kanold-McIntyre earned a bachelor of arts in elementary education from Wheaton College and a master of arts in educational leadership from Northern Illinois University. She will complete her doctorate in educational leadership at National Louis University.

To learn more about Jessica Kanold-McIntyre's work, follow her @jkanold on Twitter.

Georgina Rivera is a voice for equitable mathematics instruction. Georgina currently works as a principal in West Hartford, Connecticut. She has previously served as a middle school mathematics teacher, K–5 mathematics coach and STEM supervisor, and dean of students at a K–8 school. As a result of her mathematics leadership, the school reached higher academic achievement and growth, and as result, the school was named CAS Middle School of the Year in 2023.

Georgina is active in many professional organizations and serves on several boards including serving as the vice president of National Council of Supervisors of Mathematics. In addition to being a servant leader and volunteer, she is an author, blogger, and coach and enjoys leading professional learning.

To learn more about Georgina Rivera's work, follow her @mathcoachrivera on Twitter.

To book Sarah Schuhl, Timothy D. Kanold, Mona Toncheff, Bill Barnes, Matthew R. Larson, Jessica Kanold-McIntyre, or Georgina Rivera for professional development, contact pd@SolutionTree.com.

Preface

By Timothy D. Kanold

In the early 1990s, I had the honor of working with Richard DuFour at Adlai E. Stevenson High School in Lincolnshire, Illinois. During that time, Rick—then principal of Stevenson—began his revolutionary work as one of the architects of the Professional Learning Communities at Work® (PLCs at Work) process. My role at Stevenson was to create, initiate, and incorporate the elements of the PLC process into the preK–12 mathematics programs, including the K–5 and 6–8 schools feeding into the Stevenson district.

In those early days of building the PLC process and collaborative culture, my colleagues and I exchanged knowledge about our personal growth and improvement as teachers and began to create and enhance student agency for learning mathematics. As colleagues and team members, we taught, coached, and learned from one another. Our work became more transparent.

And yet, we did not know our mathematics *assessment* story. We did not have much clarity on a collaborative assessment vision that would significantly improve student learning and erase unintentional inequities caused by our private decision making for our mathematics assessment.

We did not initially understand how the work of our collaborative teacher teams, when focused on the right *assessment* actions, could erase inequities in student learning caused by the wide variance in the products, processes, and academic rigor of our professional assessment practices.

Through our work together, we realized that, without intending to, our isolated decisions often created the crushing consequence of poor student performance in a vertically connected curriculum like mathematics.

We also realized the benefit of belonging to something larger than ourselves. There is a benefit to learning about various teaching and assessing strategies from each other, *as professionals*. It was often in community that we found deeper meaning to our work, and strength in the journey, together.

As we began our collaborative mathematics work at Stevenson and with our feeder districts, we discovered quite a bit about our mathematics teaching by studying together our student performance data, observing one another, and learning from one another about success in our professional work.

We discovered that if we were to become what was about to be called a PLC, then experimenting together in our grade-level and course-based teams by creating, sharing, and formatively using effective assessment routines needed to become our norm. We needed to learn more about how the essential mathematics standards, the nature of the mathematics tasks we were choosing for the assessment of those standards, and the types of student action taken because of our assessment feedback could maximize student learning. And we realized we needed to become more assessment literate, *together*.

This idea of a collaborative focus on the real work we do as mathematics teachers is at the heart of the *Every*

Student Can Learn Mathematics series. The authors of this book believe that if teachers do the right work together and develop positive routines together, then every student can learn grade-level or course-based expectations in mathematics. This belief that began with the mighty team of mathematics teachers at Stevenson has been the driving force of the authors' work for more than thirty years now.

In this series, we emphasize the concept of team actions. We recognize some readers may be the only members of a grade level or mathematics course. In that case, we recommend you work with a colleague in a grade level or course above or below your own. Or work with other job-alike teachers across a geographic region as technology allows.

The collaborative team becomes the engine that drives the PLC at Work process forward with a positive and ongoing analysis of student data based on the evidence of student learning.

A PLC at Work in its truest form is "an ongoing process in which educators work collaboratively in recurring cycles of collective inquiry and action research to achieve better results for the students they serve" (DuFour, DuFour, Eaker, Many, & Mattos, 2016, p. 10). This book and the others in the *Every Student Can Learn Mathematics* series feature a wide range of voices, tools, and discussion protocols offering advice, tips, and knowledge for your PLC-based collaborative mathematics team.

The lead authors of the *Every Student Can Learn Mathematics* series—Sarah Schuhl and Mona Toncheff—have each been on their own journeys with the deep and collaborative PLC work for mathematics. They have spent significant time in the classroom as highly successful practitioners, leaders, and coaches of preK–12 mathematics teams, designing and leading the structures and the culture necessary for effective and collaborative team efforts. They, along with our additional coauthors and associates, have lived through and fiercely led the mathematics professional growth actions this book advocates within diverse preK–12 settings in rural, urban, and suburban schools.

In this book, we tell our mathematics assessment and intervention story. It is a story of developing high-quality assessments for students, scoring those assessments with fidelity, and helping students use those assessments for reflection and formative learning. It is a preK–12 story that, when well implemented, will bring great satisfaction to your work as a mathematics professional and result in a positive impact on your students.

We hope you will join us in the journey of significantly improving student learning in mathematics by leading and improving your assessment story for your team, your school, and your district. The conditions and the actions for adult learning of mathematics together are included in these pages. We attempt to guide your conversations and your assessment work with colleagues to make the journey a bit easier as you analyze data together and use the results to improve student learning and mathematics performance.

We hope the personal stories we tell, the assessment framework we provide, the tools we use, and the reflection opportunities we include serve you well in your daily work in a discipline we all love—mathematics!

Introduction

If you are a teacher of mathematics, then this book is for you! Whether you are a novice or a master teacher; an elementary, middle, or high school teacher; or a rural, suburban, or urban teacher, this book is for you. It is for all teachers and support professionals who are part of the preK–12 mathematics learning experience.

Your professional life as a mathematics teacher is not easy. In this book, you and your colleagues will focus your time and energy on collaborative assessment efforts that result in significant improvement in student learning.

Some educators may ask, "Why become engaged in collaborative mathematics teaching actions in your school or department?" The answer is simple: *equity*.

What is equity? To answer that question, it is helpful to examine inequity. In traditional schools where teachers work in isolation, there is often a wide discrepancy in teacher practice. Teachers in the same grade level or course may teach and assess mathematics quite differently—there may be a lack of consistency in what teachers expect students to know and be able to do, how they will know when students have learned, what they will do when students have not learned, and how they will proceed when students have demonstrated learning. Such wide variance in potential teacher practice among grade-level and course-based teachers causes subsequent inequities as students pass from grade to grade and course to course.

These types of equity issues require you and your colleagues to engage in team discussions around the development and use of mathematics assessments that provide evidence of and strategies for improving student learning.

Equity and PLCs

The PLC at Work process is one of the best and most promising models your school or district can use to build a more equitable response for student learning. Richard DuFour, Robert Eaker, and Rebecca DuFour, the architects of the PLC at Work process, designed the process around three big ideas and four critical questions that place learning, collaboration, and results at the forefront of our work as teachers (DuFour et al., 2016). Schools and districts that commit to the PLC transformation process rally around the following three big ideas (DuFour et al., 2016).

1. **A focus on learning:** Teachers focus on learning as the fundamental purpose of the school rather than on teaching as the fundamental purpose.

1. **A collaborative culture:** Teachers work together in teams to interdependently achieve a common goal or goals for which members are mutually accountable.

1. **A results orientation:** Team members are constantly seeking evidence of the results they desire—high levels of student learning.

Additionally, teacher teams within a PLC at Work focus on four critical questions (DuFour et al., 2016).

1. What do we want all students to know and be able to do?

2. How will we know if students learn it?

3. How will we respond when some students do not learn?
4. How will we extend the learning for students who are already proficient?

The four critical questions of a PLC provide an equity lens for your professional work. Imagine the opportunity gaps that will exist if you and your colleagues do not agree on the level of rigor for PLC critical question 1 (DuFour et al., 2016): What do we want all students to know and be able to do?

Imagine the devastating effects on preK–12 students if you do not reach complete team agreement on the high-quality criteria for the mathematics assessments you administer (PLC critical question 2) and your routines for how you score them. Imagine the lack of student agency (student voice in learning) if you do not work together to create a unified, robust formative mathematics assessment process for helping students own their responses when they are and are not learning.

To answer these four PLC critical questions well requires structure through the development of products for your work together, and a formative learning culture using a reflection and action process of how you work with your teacher team.

The concept of your team reflecting together and then taking action around the right work is emphasized in the *Every Student Can Learn Mathematics* series.

The Reflect, Refine, and Act Cycle

Figure I.1 illustrates the reflect, refine, and act cycle, our perspective about the process of lifelong learning—for us, for you, and for your students. The very nature of our profession—education—is about the development of skills toward learning. Those skills are part of an ongoing process we pursue together.

More important, the reflect, refine, and act cycle is a *formative* student learning cycle we describe throughout all books in the *Every Student Can Learn Mathematics* series. When your teacher team embraces mathematics learning as a *process*, your students reflect, refine, and act by asking the following questions.

- **Reflect:** "How well do I make sense of the mathematics task, solve it with a chosen strategy, and determine, 'Is this the best solution strategy?'"
- **Refine:** "How well do I learn mathematics based on feedback about my work, solution pathways, and any possible errors?"
- **Act:** "How well do I persevere, apply learning from the mathematics task to future tasks, and determine what have I learned that I can use again?"

The intent of this *Every Student Can Learn Mathematics* series is to provide you with a systemic way to structure and facilitate the deep team discussions necessary to lead an effective and ongoing adult and student learning process each and every school year.

Mathematics in a PLC at Work Framework

The *Every Student Can Learn Mathematics* series includes three books that focus on a total of five teacher team actions within three larger categories.

1. *Mathematics Assessment and Intervention in a PLC at Work, Second Edition*
2. *Mathematics Instruction and Tasks in a PLC at Work, Second Edition*
3. *Mathematics Homework and Grading in a PLC at Work*

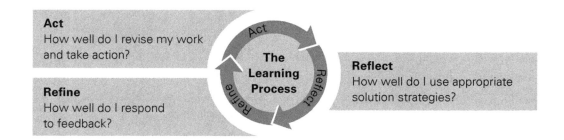

Figure I.1: Reflect, refine, and act cycle for formative student learning.

Figure I.2 shows each of these three categories and the teacher team actions within each. These five team actions focus on the nature of the ongoing, unit-by-unit professional work of your teacher team and how you should respond to the four critical questions of a PLC at Work (DuFour et al., 2016).

Most commonly, a collaborative team consists of two or more teachers who teach the same grade level or course. Through your focused work addressing the four critical questions of a PLC at Work (DuFour et al., 2016), you provide every student in your grade level or course with equitable learning experiences and expectations, opportunities for sustained perseverance, and robust formative feedback, regardless of the teacher they receive.

If, however, you are a singleton (a lone teacher who does not have a colleague who teaches the same grade level or course), you will have to determine who it makes the most sense for you to work with as you strengthen your lesson design and student feedback skills. Leadership consultant and author Aaron Hansen (2015) and consultants and authors Brig Leane and Jon Yost (2022) suggest the following possibilities for creating teams for singletons.

- Vertical teams (for example, a primary school team of grades K–2 teachers or a middle school mathematics department team for grades 6–8)
- Virtual teams (for example, a team comprising teachers from different sites who teach the same grade level or course and collaborate virtually with one another across geographic regions)
- Grade-level or course-based teams (for example, a team where teachers expand to teach and share two or three grade levels or courses instead of each only teaching all sections of one grade level or course)

Every Student Can Learn Mathematics series' Team Actions Serving the Four Critical Questions of a PLC at Work	1. What do we want all students to know and be able to do?	2. How will we know if students learn it?	3. How will we respond when some students do not learn?	4. How will we extend the learning for students who are already proficient?
Mathematics Assessment and Intervention in a PLC at Work, Second Edition				
Team action 1: Develop high-quality common assessments for the agreed-on essential learning standards.	■	■		
Team action 2: Analyze and use common assessments for formative student learning and intervention.			■	■
Mathematics Instruction and Tasks in a PLC at Work, Second Edition				
Team action 3: Develop high-quality mathematics lessons for daily instruction.	■	■		
Team action 4: Analyze and use effective lesson design elements to provide formative feedback and build student perseverance.			■	■
Mathematics Homework and Grading in a PLC at Work				
Team action 5: Develop and use high-quality common grading components and formative grading routines.	■	■	■	■

Figure I.2: Mathematics in a PLC at Work framework.

Visit **go.SolutionTree.com/MathematicsatWork** *for a free reproducible version of this figure.*

About This Book

Every grade-level or course-based collaborative team in a PLC culture is expected to meet on an ongoing basis to discuss how its mathematics lessons and assessments ask and answer the four critical questions as students are learning (DuFour et al., 2016). For this book in the series, we explore two specific *assessment team actions* for the professional work of your collaborative team.

- **Team action 1:** Develop high-quality common assessments for the agreed-on essential learning standards.
- **Team action 2:** Analyze and use common assessments for formative student learning and intervention.

This book emphasizes intentional differentiation between the mathematics assessment instruments or *products* your teacher team develops and produces (team action 1) and the *process* for how your team uses those products to analyze the student learning results of your mathematics work week after week (team action 2).

- The first chapter addresses the importance of your teacher team's use of common assessments for formative student and teacher learning. You will find a teacher team rubric with six criteria to analyze how well your team uses common assessments. The common assessment formative process rubric frames the remaining chapters in the book.
- Chapters 2 and 3 examine the quality of the common mathematics assessment instruments you and your team use during and at the end of each unit of study (team action 1). You will evaluate your current common assessments (quizzes and tests) based on eight mathematics assessment design criteria, including the calibration of your scoring for those assessments.
- Chapters 4, 5, and 6 explore how to use the high-quality common assessment instruments you develop to enhance the student and teacher learning process (team action 2). What are teachers doing with the results? What are students doing with the results? Chapter 4 shares data-analysis protocols for your teacher team to use when analyzing student work to determine next instructional steps and follow the calibration work discussed in chapter 3. Chapter 5 explores how students use assessments as part of their learning story through intentional team reflection tools and plans. Students use the feedback from their common assessments to analyze their own learning. Chapter 6 addresses how your teacher team creates effective Tier 2 intervention learning experiences for students who need additional time and support to learn targeted skills.

The tools and protocols in this book are designed to help you become confident and comfortable in mathematics assessment conversations with one another, and move toward greater transparency in your assessment practices with colleagues. Visit **go.SolutionTree.com/ MathematicsatWork** for free reproducible versions of tools and protocols that appear in this book and for additional examples to those provided in the chapters. A QR code appears throughout this book that leads to the reproducibles and additional materials.

In this book, you will find discussion tools that offer questions for reflection, discussion, and action on your teacher team mathematics assessment practices and routines. We invite you to write your personal assessment story in the teacher reflection boxes as well. You will also find some personal stories from the authors of this series. These stories provide a glimpse into the authors' personal insights, experiences, and practical advice connected to some of the strategies and ideas in this book.

The *Every Student Can Learn Mathematics* series is steeped in the belief that your decisions and your daily actions matter. As a classroom teacher of mathematics, you have the power to choose the mathematical tasks students are required to complete during lessons, homework, unit assessments such as quizzes and tests, and projects and other high-performance tasks you design. You have the power to determine the rigor for those mathematical tasks, the nature of student communication and discourse to learn those tasks, and whether or not learning mathematics should be a formative process for you and your students.

As you embrace the belief that together, you and your colleagues can overcome the complexities of your work and the learning obstacles you face each day in your professional learning community, then *every student can learn mathematics* just may become a reality in your school.

CHAPTER 1

The Mathematics at Work™ Common Assessment Process

> Assessment is a process that should help students become better judges of their own work, assist them in recognizing high-quality work when they produce it, and support them in using evidence to advance their own learning.
>
> —*NCTM, Principles to Actions*

Chances are, you already use mid-unit and end-of-unit mathematics assessments to measure and record student learning. You may or may not share those assessments with your colleagues, and for any common assessments you do give, you may or may not know the routines your colleagues use to ensure the assessments you create are fully in common. You also may not yet have a common strategy for assessing the quality of those assessments or using the assessment feedback to advance your daily learning and the learning of your colleagues and your students.

In *Learning by Doing*, authors Richard DuFour, Rebecca DuFour, Robert Eaker, Thomas W. Many, and Mike Mattos (2016) indicate that common assessments:

- Promote efficiency for teachers
- Promote equity for students
- Provide an effective strategy for determining whether the guaranteed curriculum is being taught and, more importantly, learned
- Inform the practice of individual teachers
- Build a team's capacity to improve its program
- Facilitate a systematic, collective response to students who are experiencing difficulty
- Offer the most powerful tool for changing the adult behavior and practice (p. 149)

Reflect for a moment about the students' point of view regarding the mathematics assessments they receive throughout the school year. In a PLC culture, the student mathematics assessment experience should be the same, regardless of the teacher the school assigns. The rigor of the mathematical tasks for each essential learning standard, the format of the test questions, the number of test questions, the alignment of those questions to each standard, the strategies students use to demonstrate learning, and the scoring of those assessment questions all depend on your willingness to work together and erase any potential inequities that a lack of transparency with colleagues can cause.

Working together also means creating common mathematics assessments so students experience the same mathematical rigor expectations across classrooms. In a vertically connected curriculum like mathematics, the equitable use of common unit assessments is more likely to affect the student preK–12 learning experience. The purposes behind common assessments are significant.

Four Specific Purposes

There are four specific purposes supporting the use of common mathematics assessments.

1. Improved equity and opportunity for student learning
2. Professional development
3. Student ownership of learning
4. Student intervention and support from the teacher team

Improved Equity and Opportunity for Student Learning

Essentially, you minimize the wide variance in student expectations between you and your colleagues (an inequity creator) when you work collaboratively to design high-quality assessment instruments appropriate to the identified essential learning standards for the unit.

Common mathematics unit assessments, then, are the first order of business for your professional work life. They are at the heart of working effectively and more equitably as your collaborative team answers the second critical question of the PLC culture (DuFour et al., 2016): How will we know if students learn it? Agreement on the mathematical tasks used to assess each essential learning standard for a unit of study is where team collaboration often begins.

Professional Development

When working collaboratively to design high-quality mathematics assessments, you learn from colleagues and gain clarity as to the rigor of the tasks for each essential standard of a mathematics unit. This expectation often requires team discussion on the cognitive-demand level for each mathematics task or problem on the assessment and the appropriate amount of application or mathematical modeling. As your team asks, "Do the tasks (the assessment questions) align to the expectations of the learning standards?" you begin a professional development process about the mathematics content and the strategies to teach and assess the content.

You build the capacity of your teacher team to learn from one another as you share insights into the common assessment tasks and expectations for students to receive full credit in response to any mathematics problem on the assessment.

TEACHER *Reflection*

How do you currently determine if the quality of the mathematical tasks (the test questions) you use to assess student learning is appropriate to the understanding levels expected by each essential learning standard of the unit?

_____
_____
_____
_____
_____
_____
_____
_____
_____

Personal Story **SARAH SCHUHL**

When I first began working as part of a professional learning community at Centennial High School, we began our assessment work as a team in geometry. We discovered some teachers heavily weighted proofs and required students to solve quadratic equations when performing mathematical tasks related to standards for specific angles.

Other teachers on our team only required fill-in-the-blank proofs using less complex linear equations for the same standards. This meant we were using variant levels of rigor for the same standard, and equitable learning experiences and expectations for our students did not exist in geometry.

Similarly, when I worked as an instructional coach and looked at an assessment related to rounding with my fourth-grade colleagues, we noticed that some of the teachers only asked students to round to the nearest ten, others required explanations, and still others asked questions for conceptual understanding, such as, What is the largest number that could be rounded to fifty when rounding to the nearest ten?

In addition, you and your colleagues benefit from your analysis of student learning evidence from your common mid- and end-of-unit assessments. Together, you identify which essential standards your students learned or did not yet learn, determine individual student learning needs, and clarify trends in student thinking and understanding. You and your team can use the data to analyze the effectiveness of your instructional practices and together design targeted interventions and extensions students might need during and after the mathematics unit.

Student Ownership of Learning

As a result of common mathematics assessments, you and your colleagues can help students identify which mathematics standards they have learned or not yet learned and work together to find appropriate and common responses to interventions or extensions (discussed in chapter 6 of this book). Your students begin to take greater ownership in their learning of the essential mathematics standards as they respond to evidence of learning from the common unit assessments.

As your students learn to use daily student trackers to monitor their progress on each essential learning standard, they can begin to understand or view mathematics assessment not as something to be feared, but rather as an integral aspect of their learning process. Consider how students develop a positive and productive mathematics identity through self-reflection and ownership of learning. How do students see assessments as part of their learning story instead of their judging, measuring, and recording story? You can find sample student trackers in chapter 5 on pages 102–104.

Student Intervention and Support From the Teacher Team

Student support systems for mathematics should rely on the essential learning standards for the unit. When mid- and end-of-unit assessments are common, it allows each member of your team to help any students in the grade level or course you teach. If a student is not sure about a specific learning standard and needs help, the student can receive help from any teacher on your team, and each teacher can anticipate the usual trouble spots that may occur. The intentional and targeted mathematics intervention (discussed in chapter 6, page 121) creates more unified, engaging, and equitable learning experiences for students.

TEAM RECOMMENDATION

Know the Purpose of Common Assessments

- Understand how common assessments erase inequities in student learning and allow for teacher learning, provide opportunities for students to identify what they have learned and not learned, and develop a collective response to re-engaging students in learning.

- Make a team commitment to the development of common end-of-unit and mid-unit assessments throughout the year.

The purposes of common assessments reveal that your team's actions toward creating a common mathematics assessment tool are a necessary part of your professional work, but not sufficient to fully impact student learning. Those actions are only the beginning of a full formative feedback and assessment process for you, your colleagues, and your students.

The Assessment Instrument Versus the Assessment Process

What happens in the classrooms across your collaborative team when you return a common mathematics assessment to your students? What do you expect students to do with the feedback you provide? How does your collaborative team analyze data and learn from the common assessments you give and the feedback you provide? How will your students take action on their errors? The answers to these questions inform your formative assessment process.

Not only should you and your colleagues create the common assessment instrument itself (the actual quiz or test for each unit), but you also should establish a process in which you and your colleagues use this assessment with students. This distinguishes between the formative assessment process you use and the assessment instruments you design as part of the formative learning process. DuFour and colleagues (2016) describe the importance of having common instruments and a common process this way:

> One of the most powerful, high-leverage strategies for improving student learning available to schools is the creation of frequent, high-quality common formative assessments by teachers who are working collaboratively to help a group of students acquire agreed-on knowledge and skills. (p. 141)

W. James Popham (2011) provides an analogy to describe the difference between summative assessment instruments (such as your end-of-unit tests) and formative assessment processes (such as what your students do in each lesson to learn the standards or what your students do as a response to their test results). Though over a decade old at the time of this second edition's publication, his analogy still stands. He describes the difference between a surfboard and surfing (Popham, 2011):

> While a surfboard represents an important tool in surfing, it is only that—a part of the surfing process. The entire process involves the surfer paddling out to an appropriate offshore location, selecting the right wave, choosing the most propitious moment to catch the chosen wave, standing upright on the surfboard, and staying upright while a curling wave rumbles toward shore. (p. 36)

The surfboard (the test) is a key component of the surfing process, but it is not the entire process. Sarah describes her initial experience with assessments as an example of the need for the full surfing process at the bottom of this page.

Sarah's story illustrates a statement that appears in NCTM's (2014) *Principles to Actions*: "Thinking of assessment as limited to 'testing' student learning rather than as a process that can advance it has been an obstacle to the effective use of assessment processes for decades" (p. 91). The mathematics assessment instruments your team creates are the tools your team uses to collect data about student demonstrations of the essential learning standards. The assessment instruments subsequently inform your and your students' ongoing decisions about learning mathematics throughout the school year. Jan Chappuis and Rick Stiggins (2020) state that you and your colleagues should use assessments to "do two things: (1) gather accurate information about student achievement and (2) use the assessment process and its results effectively to improve achievement" (p. 2).

Assessment instruments vary and can include tools such as class assignments, observed work at stations, exit slips, journal entries, quizzes, unit tests, projects, and performance tasks, to name a few. How each assessment instrument is given also varies and may include using paper and a pencil, using a computer or

---

Personal Story SARAH SCHUHL

Not all teaching experiences are the same. My first mathematics teaching job was at a small rural school serving grades 7–12 in eastern Oregon. Not only was I the mathematics teacher, I was the mathematics department. As such, I dutifully created and scored mid-unit quizzes and chapter tests for each course—just as I had experienced as a student many years before. Occasionally, I wondered if the assessments were good only to quickly convince myself that since I had taken questions from the publisher resources, they must be fine.

Fast-forward to my second mathematics teaching position outside of Portland, Oregon, where teachers worked in teams and gave the same common assessments—mid-unit checks and end-of-unit tests. This was an eye-opening experience for me. I quickly learned the efficiency of collaboratively planning an assessment to make clear the learning expectations in each unit and ensure the results matched the learning expectations for student instruction. Incredible! Who knew that creating a common unit assessment could impact and grow student learning, not just measure and record it? This in turn grew my own understanding of the content, which then strengthened my daily instruction.

tablet, or being observed in real time. However, to avoid inequities in the level of rigor for student work, and to serve the formative learning process, these assessment instruments must be in common for every teacher on your team. How, then, can your team design and use assessments to enhance and promote student learning? In this book, we seek to support your collaborative team in answering this essential question.

Richard DuFour (2015) states:

> When members of a team work together to consider how they will approach assessing the skills and knowledge of their students, when every item or performance task is vetted by the entire team before it becomes part of the assessment, there is a much greater likelihood that the end result is a quality assessment compared to assessments that isolated teachers create. (p. 175)

In addition, there are three important elements to consider for using common mathematics assessment results effectively. First, your team needs to collectively analyze the data and student work from common assessments to evaluate instructional practices and create targeted intervention or extension learning experiences.

Second, students need to receive *meaningful assessment feedback* on their work in order to improve their mathematical understanding. For the purposes of this book, meaningful assessment feedback refers to the FAST feedback elements described in this chapter.

Third, students need to use tools or protocols requiring them to take action on mid-unit and end-of-unit assessment feedback using a formative process.

FAST Feedback

The Mathematics in a PLC at Work team uses the acronym *FAST* (*fair, accurate, specific,* and *timely*) to describe the essential characteristics of meaningful formative feedback explained by Douglas Reeves (2011, 2016) and John Hattie (2009, 2012; Hattie, Fisher, & Frey, 2017). FAST feedback works well when using evidence of learning from common assessments during or at the end of the unit. This feedback includes the following characteristics.

- **Fair:** Effective feedback on the assessment rests solely on the quality of the student's demonstrated work and not on other characteristics.

- **Accurate:** Effective feedback on the mathematics assessment acknowledges what students are currently doing well and correctly identifies errors they are making. According to Jan Chappuis and Rick Stiggins (2020), "descriptive feedback should reflect strengths and weaknesses with respect to the specific learning target(s) [students] are trying to hit" (p. 36).

- **Specific:** Your mathematics assessment notes and feedback "should be about the particular qualities of [the student's] work, with advice on what [the student] can do to improve, and should avoid comparison with other pupils" (Black & Wiliam, 2001, p. 6). Try to find the right balance between being specific enough so the student can quickly identify the error in their logic or reasoning and not being so specific that you end up correcting the work for them. Does your assessment feedback help students correct their thinking as needed?

- **Timely:** You provide effective feedback on the end-of-unit mathematics assessment just in time for students to take formative learning action on the results before too much of the next unit takes place. As a general rule, you should pass back to students mid-unit and end-of-unit assessments with expected results for proficiency within forty-eight hours of the assessment.

Kim Bailey and Chris Jakicic (2019) share, "We see assessment as something we do with students rather than something we do to students" (p. 122). Dylan Wiliam (2016) states if your feedback doesn't *grow the learning* of students and change or validate their understanding, then it has been a waste of time. What students *do* with your FAST feedback is important.

What your teacher team learns about student thinking from the common assessment results creates additional targeted student learning opportunities from the common assessment feedback students receive. This allows you to use your team's assessment results as part of a *formative* learning process. The assessment becomes much more than an *event* for students. It becomes a valuable tool in the learning *process*.

Use the discussion tool in figure 1.1 (page 10) to determine your team's current reality related to using high-quality FAST feedback with your common assessments.

Directions: Use the following prompts to guide team discussion of your current use of FAST feedback for all common during-the-unit and end-of-unit mathematics assessments.

Feedback to students:

1. How quickly do students receive their assessment results from each teacher on the team?

2. How do you provide corrective feedback to student errors on the common assessment questions?

Student actions:

3. What do you expect students to do with the assessment when you return the assessment?

4. How do you require students to reflect on their learning based on evidence from assessment results?

Teacher intervention actions:

5. How does your teacher team analyze trends in student work (for example, common misconceptions) when discussing common assessment results? What do you do with that information?

6. What does your teacher team collectively do to help students re-engage in learning content?

Figure 1.1: Teacher team discussion tool—Formative use of common assessments.

*Visit **go.SolutionTree.com/MathematicsatWork** for a free reproducible version of this figure.*

Evaluation of Mathematics Assessment Feedback Processes

Throughout the feedback process, you and your colleagues should ask students to reflect on the evidence of their learning, refine their understanding of mathematical concepts and applications, and act on that knowledge to demonstrate proficiency. Feedback is only effective if students act on it and take ownership of their learning. Along the way, you and your students use the assessment instruments as learning tools within a formative feedback loop.

The personal story at the bottom of the page from author Jessica Kanold-McIntyre highlights the benefits of using a team-developed student feedback process.

You can use figure 1.2 (page 12) to identify areas to improve in your team's formative assessment process. The action students take on your FAST feedback and your coordinated team efforts at designing equitable quality interventions impact every student. Each of the criteria is explained in greater detail in chapters 2–6.

Use figure 1.2 to evaluate your current team effort to use common assessments as part of a meaningful formative feedback process with students. Give your teacher team a score of 1 to 4 for each of the six criteria.

After self-evaluating your current progress on the common unit assessment criteria, what did your team discover are areas of strength and areas in need of improvement?

Formative feedback for students is most effective when carefully planned. Intentional planning includes determining the essential learning standards and creating common unit assessments that reveal student thinking and learning. From the data revealed through student work on the assessments, your team can plan for student reflection and goal setting for continued learning. You can also plan for students to re-engage in learning through intervention opportunities your team provides.

TEACHER *Reflection*

How did you score your grade-level or course-based mathematics team on the six criteria from figure 1.2 (p. 12)? How did your team do? Describe the most urgent assessment issue you need to address with your colleagues related to your formative use of common assessments.

_____
_____
_____
_____
_____
_____
_____
_____
_____
_____
_____
_____

Personal Story **JESSICA KANOLD-MCINTYRE**

There is a great power in the calibration process! I worked with a first-grade team that had been giving common assessments for a while. However, the teachers had just started talking about how they gave feedback on assessments and how students reflected on their assessments by learning target. The team had developed a feedback form for students to use when they passed back the common mathematics assessments. The teachers were so excited to use the feedback form with their students.

When asked how the feedback form was working, the five teachers on the team realized that they all used the form in a different way. Until that moment when they heard each other share their expectations for student responses to errors on each standard, they had not realized they each used the form differently and at different times. The teachers had all assumed everyone interpreted the expectations similarly and were surprised by the unintentional inequities they were causing their students. The conversation became an amazing collaboration experience in which they worked together to come to consensus on each part of FAST feedback and created a team Tier 2 intervention response to the assessment they could all share. Their willingness to discuss how to use the form to best support student learning required each of them to also approach the conversation in a professional and respectful manner.

Team Common Assessment Formative Process Criteria	Description of Level 1	Requirements of the Indicator Are Not Present	Limited Requirements of the Indicator Are Present	Substantially Meets the Requirements of the Indicator	Fully Achieves the Requirements of the Indicator	Description of Level 4
1. Agreed-on essential learning standards for the unit	Essential learning standards for each unit are unclear or differ among teachers on the team.	1	2	3	4	Teachers on the team ensure the learning standards are clear and commonly worded for students and shared with students at the start and throughout each unit.
2. High-quality common unit assessments	Teachers on the team use their own mid-unit assessments and end-of-unit assessments. They do not share those assessments with colleagues.	1	2	3	4	Teachers on the team design and administer high-quality common unit assessments using the eight criteria of assessment design (figure 2.1), coordinate when and how to give the assessments, and collectively respond to student learning progress and data.
3. Calibration routines	Teachers on the team score assessments individually and do not share the nature of the feedback they provide to students.	1	2	3	4	Teachers on the team regularly calibrate their scoring of common assessments to verify accurate scoring of student work and determine the best way to provide high-quality actionable feedback to students.
4. Teacher data-analysis and action routines	Teachers individually analyze data from assessments for the primary purpose of assigning grades and identify students who learn or do not learn.	1	2	3	4	Teachers on the team analyze data (including student work) from common assessments to identify effective instructional practices, determine levels of student mastery for each essential learning standard, reveal trends in student thinking, and implement targeted interventions and extensions.
5. Student self-assessment and action routines	Teachers do not ensure students reflect on learning. Common assessments are not used to identify standards students have learned and to help students plan for future learning.	1	2	3	4	Teachers on the team create a system that includes a tool for students to self-assess what they learned or did not learn by essential learning standard with a plan to re-engage in learning within team-created intervention structures.
6. Team response to student learning using Tier 2 intervention criteria	Teachers on the team each determine how they provide students opportunities for intervention. Teachers on the team design interventions independently of one another.	1	2	3	4	Teachers on the team develop a collective, just-in-time, targeted response to student learning by student and by essential learning standard, creating structures and plans for students to re-engage in and demonstrate learning as a result of teacher team actions.

Figure 1.2: Teacher team rubric—Common assessment formative process.

Visit go.SolutionTree.com/MathematicsatWork for a free reproducible version of this figure.

Reflection on Team Assessment Practices

In order to create high-quality common mathematics assessment instruments, you and your colleagues first need to understand the purposes of the assessments you administer to students. The questions in figure 1.3 (page 14) will help you and your team understand one another's perspectives related to common assessments, particularly about the design, value, and uses of common assessments related to student learning, and you'll begin to frame the work ahead in this book.

Use the self-reflection protocol in figure 1.3 to ask each member of your mathematics team about their personal unit-by-unit assessment practices. You can use team members' responses to find some initial common ground as you share your current mathematics assessment work.

Discuss your responses as a collaborative team to determine how similar or different your practices are as you build your team consensus for common assessment development.

Used well, common assessment instruments (your surfboards, so to speak) provide quality direction to you, your colleagues, and your students. The design of your unit assessments should also support your team's important assessment work. Consider your team response to the first two critical questions of a PLC at Work (DuFour et al., 2016).

1. What do we want all students to know and be able to do?
2. How will we know if students learn it?

A proper team response to these two questions will be revealed through the mathematics assessments you and your team design for each unit throughout the school year. This is **team action 1:** develop high-quality common assessments for the agreed-on essential learning standards.

TEACHER *Reflection*

Reflect with your colleagues on the current process you use to design unit assessments.

1. Who writes the assessments: each teacher on the team, the central office, authors of published resources you've purchased, or some other entity?
2. What sources do you use?
3. How do you make decisions about scoring each mathematical task or question on the assessment?

Directions: Use the following prompts to guide team discussion of your common assessment practices for each unit. Share your results with other teams as needed.

Design of common mathematics assessments:

1. How do we organize our common mathematics assessments (for example, by assessment type, by rigor of the problems, by essential learning standard, or in another way)?

2. Do we expect students to show their work on each common mathematics assessment? What work do we expect?

3. Are there common scoring agreements for each mathematics task (problem), or each essential learning standard, on the common mathematics assessment?

Use of common mathematics assessments:

4. How do we currently analyze data from our common mathematics assessments? How do we use the data to inform our instructional strategies and decisions for future teaching of the standards being assessed?

5. How do we use our assessment data to provide for targeted team interventions and to influence our teaching in the next unit of mathematics?

6. How are grades and comments on assessments used to provide feedback to students about their learning and errors?

7. How are students currently required to respond to errors when receiving timely feedback and scoring on their common assessments?

Figure 1.3: Teacher team discussion tool—Common assessment self-reflection protocol.

*Visit **go.SolutionTree.com/MathematicsatWork** for a free reproducible version of this figure.*

Additionally, it is only through the nature of the expected student response to your feedback during and at the end of the unit that your common assessments become formative for students (process of surfing). And it is in the formative feedback process that your intervention work as a teacher team resides. Through the formative assessment process, you and your colleagues embrace a high-quality intervention response to the third and fourth PLC critical questions (DuFour et al., 2016).

3. How will we respond when some students do not learn?
4. How will we extend the learning for students who are already proficient?

In short, how do you, your team, and your students use assessment results and evidence from student work to intervene and continue learning the essential standards for each mathematics unit? This is **team action 2:** analyze and use common assessments for formative student learning and intervention.

Mathematics assessment throughout the year becomes an opportunity for students to reflect, refine, and act (see page 2 in the introduction). Your students should be expected to reflect on their learning during the unit to choose effective strategies to solve problems, and after each assessment, the students should be expected to learn from any errors and revise their learning using feedback so that they can act on that learning in the future. Your teacher team learns about student thinking to inform next instructional steps.

Before this formative process can begin, the first order of business for any collaborative team is to evaluate the quality of its current school- or district-driven mathematics assessments as shown in the teacher team rubric in figure 1.2 (page 12).

Visit **go.SolutionTree.com/MathematicsatWork** for free reproducible versions of tools and protocols that appear in this book, as well as additional online only materials.

CHAPTER 2

Quality Common Mathematics Assessments

Design is a funny word. Some people think design means how it looks.
But of course, if you dig deeper, it's really how it works.

—Steve Jobs

How do you decide if the common unit-by-unit mathematics assessment instruments you design are high quality? This is a question every teacher and leader of mathematics should ask. Tim Kanold describes his experience at Stevenson High School District 125 when he first asked this question of the middle school mathematics teachers from one of the Stevenson feeder districts in the following personal story.

The mathematics assessment work Tim describes created an early beta model for a test evaluation tool he and his fellow teachers could use to evaluate the quality of their unit-by-unit mathematics assessments. Figure 2.1 (page 18) provides a much deeper and robust during-the-unit or end-of-unit assessment instrument evaluation tool your collaborative team can use to evaluate quality and build new and revised unit assessment instruments. Figure 2.2 (page 19) provides discussion questions to evaluate your team's application of each criterion on the assessment instrument quality evaluation rubric in figure 2.1 and includes some considerations for preK and the primary grades.

Your collaborative team should rate and evaluate the quality of its most recent end-of-unit or chapter assessment instruments using the tools in figure 2.1 and figure 2.2. How does your team's end-of-unit assessment score when you sum up the ratings? Do you rate your current assessment a 12, 16, or 22? How close does your assessment instrument come to scoring twenty-seven or higher out of the thirty-two points possible in the assessment quality protocol?

Personal Story **TIMOTHY KANOLD**

During my fourth year as the director of mathematics at Stevenson, it was clear to me that we were not very assessment literate. By this I mean we had very little knowledge about each other's assessment routines and practices. I remember collecting all the mathematics tests being given to our grades 6–8 feeder district students during the months of October and November, and feeling a certain amount of dismay at the wide range of rigor, the lack of direction for how the assessments were organized, and our lack of clarity about the nature of the mathematics tasks being chosen for the exams.

In some instances, the questions we were asking sixth-grade students were more rigorous than the questions we were asking eighth-grade students. We had little or no shared direction for this most important aspect of our professional life. We needed to do a deep dive into the research and the expectations of high-quality mathematics assessments. It began a journey for all of our mathematics teachers to not only become assessment literate, but also use our mathematics assessments as a way to sustain student effort and learning at the right level of rigor across the grades.

High-Quality Assessment Criteria	Description of Level 1	Requirements of the Indicator Are Not Present	Limited Requirements of the Indicator Are Present	Substantially Meets the Requirements of the Indicator	Fully Achieves the Requirements of the Indicator	Description of Level 4
1. Identification of and emphasis on essential learning standards (student-friendly language)	Essential learning standards are unclear, absent from the assessment instrument, or both. Some of the assessment tasks (questions) may not align to the essential learning standards of the unit. The organization of assessment tasks is not clear.	1	2	3	4	Essential learning standards are clear and written on the assessment. The assessment tasks (questions) are organized and aligned to the essential learning standards of the unit.
2. Balance of higher- and lower-level-cognitive-demand mathematical tasks	Emphasis is on procedural knowledge with minimal higher-level-cognitive-demand mathematical tasks for demonstration of understanding.	1	2	3	4	Test is rigor balanced with higher- and lower-level-cognitive-demand mathematical tasks present and aligned to the essential learning standards.
3. Variety of assessment-task formats	Assessment contains only one type of questioning strategy—selected response or constructed response. There is little to no modeling of mathematics.	1	2	3	4	Assessment includes a blend of assessment types and mathematical modeling, with opportunities for students to show their thinking in various ways.
4. Appropriate and clear scoring agreements (scoring guide or proficiency rubric)	Team scoring agreements are not evident or are inappropriate for the assessment tasks.	1	2	3	4	Scoring agreements are clearly defined and appropriate for each mathematical task or essential learning standard (scoring guide or proficiency rubric).
5. Clarity of directions	Directions are missing or unclear. Directions are confusing for students.	1	2	3	4	Directions are appropriate and clear.
6. Academic language	Wording is vague or misleading. Academic language (vocabulary and notations) is not precise, causing a struggle for student understanding and access.	1	2	3	4	Academic language (vocabulary and notations) in tasks is direct, fair, accessible, and clearly understood by students. Teachers expect students to attend to precision in response.
7. Visual presentation	Assessment instrument is sloppy, disorganized, and difficult to read, and it offers no room for student work.	1	2	3	4	Assessment is neat, organized, easy to read, and well spaced, with room for student work. There is also room for teacher feedback.
8. Logistics	Each teacher independently decides the technology, tools, and resources students can use during the assessment, as well as what scaffolding support to give, and how to respond if students do not finish in the designated time determined by the team (with the exception of predetermined individual student accommodations and modifications).	1	2	3	4	Teachers on the team agree on the tools and resources students can use during the assessment, conditions for scaffolding support, and how to respond if students do not finish in the designated time determined by the team (with the exception of predetermined individual student accommodations and modifications).

Figure 2.1: Teacher team rubric—Assessment instrument quality evaluation.

Visit go.SolutionTree.com/MathematicsatWork for a free reproducible version of this figure.

Directions: Examine your most recent common end-of-unit assessment, and evaluate its quality against the following eight criteria (from figure 2.1). Write your responses to each question in the following spaces.

1. Are the essential learning standards written on the assessment?

Discuss: What do our students think about learning mathematics? Do they think learning mathematics is about doing a bunch of random mathematics problems? Or can they explain the essential learning standards in student-friendly *I can* statements for each group of questions? Can they solve any mathematical tasks that might reflect a demonstration of learning the standards? Are our students able to use the essential learning standards and tasks to determine what they have learned and what they have not learned yet?

Note: In order for students to respond to the end-of-unit assessment feedback when the teacher passes it back, this is a necessary assessment feature. If students are taking an assessment that is randomized on the computer, consider how to tag each assessment item with the essential learning standard for team and student feedback. For preK and the primary grades, when giving one-on-one or small-group assessments, include the *I can* statement or statements in the assessment directions and at the top of any recording sheet.

2. Is there an appropriate balance of higher- and lower-level-cognitive-demand mathematical tasks on the assessment?

Discuss: What percentage of all tasks or problems on the assessment instrument are of lower-level cognitive demand? What percentage are of higher-level cognitive demand? Is there an appropriate balance? Is balancing rigor a major focus of our work? What level of cognitive demand is needed for each essential learning standard in this unit?

Note: Use figure 2.4 (page 27) as a tool to determine the level of cognitive demand needed for assessment tasks. Also, see the appendix (page 137) for more advice on this criterion. As a good rule of thumb, the assessment's rigor-balance ratio should be about 60/40 (lower- to higher-level cognitive demand) as appropriate to the standards on the assessment. For preK and kindergarten, many tasks will only be lower level to match the intent of the standards.

3. Is there a variety of assessment formats?

Discuss: Does our assessment use a blend of assessment formats or types? If we use multiple choice, do we include questions with multiple possible answers? Do we provide tasks that assess mathematical modeling? What is the expectation for students to show their thinking when answering questions? Do students have the option of showing a variety of strategies to solve tasks?

Note: Your common end-of-unit assessments should not be of either extreme—all multiple-choice or all constructed-response questions. For preK and the primary grades, team common assessments will be more constructed response (though given one-on-one or in small groups orally) and there is not a need to balance the assessment formats.

4. Are scoring agreements clear and appropriate?

Discuss: Are the scoring agreements to be used for every task (scoring guide) or for each essential learning standard (proficiency rubric) clearly stated on the assessment? Do our scoring agreements make sense based on the complexity of reasoning for the tasks? Is there clear understanding of the student work necessary to receive full credit for each assessment task or question? Is it clear to each team member how partial credit will be assigned?

continued →

Figure 2.2: Teacher team discussion tool—High-quality assessment evaluation.

5. **Are the directions clear?**

   Discuss: What does clarity mean to each member of our team? Are any of the directions we provide for the different assessment tasks confusing to the student? Why?

   Note: The verbs (action words) you use in the directions for each set of tasks or problems are very important to notice when discussing clarity. Also, be sure that in the directions, you clearly state the student work you expect to see and will grade using points or a scale. If tasks are multiple choice, be sure to clarify if students should select one answer or "all that apply."

6. **Is the academic language precise and accessible?**

   Discuss: Are the vocabulary and notations for each task we use on our common assessment clear, accessible, and direct for students? Do we attend to the precision of language used during the unit, and do students understand the language we use on the assessment? How will our students be expected to demonstrate understanding as it relates to precise academic language and notations for the assessment?

   Note: The assessment instrument should include the proper language supports for all students.

7. **Does the visual presentation allow students to show what they have learned?**

   Discuss: Do our students have plenty of space to write out solution pathways, show their work, and explain their thinking for each task on the assessment instrument? Are any coordinate planes on the assessment large enough for students to accurately use when drawing a graph? Is the font size large enough for students to clearly read?

   Note: This criterion often is one of the reasons not to use the written tests that come with your textbook series. You can use questions from the test bank aligned to your instruction, but space mathematics tasks and assessment questions as needed to allow plenty of room for students to demonstrate their understanding. For preK and the primary grades, when using one-on-one or small-group assessments, make student assessment papers visually inviting and any flash cards or manipulatives clear to read and use.

8. **Are the logistics of the assessment clear and agreed on?**

   Discuss: How long should it take students to complete this assessment? What will be our procedure if they cannot complete the assessment within the allotted time so all students receive equitable opportunities to demonstrate learning? What technology, tools, and resources can be available for student use? Which students, if any, can receive scaffolding supports (in addition to students with predetermined accommodations or modifications)?

   Note: Each teacher on the team should complete a full solution key for the assessment the teacher expects of students. For upper-level students, it works well to use a time ratio of three to one (or four to one) for student to teacher completion time to estimate how long it will take students to complete an assessment. For elementary students, it may take much longer to complete the assessment. All teachers should use the agreed-on time allotment. Discuss as a team the manipulatives and technology students can have access to as well as any supports that might be on the walls or on their desks from learning during the unit. Clarify how to support students also learning English or students needing predetermined accommodations or modifications.

Visit go.SolutionTree.com/MathematicsatWork for a free reproducible version of this figure.

TEACHER *Reflection*

Based on the criteria from figures 2.1 and 2.2, summarize the result of your evaluation of an assessment you created and used with your students.

What action can you immediately take to improve your assessments?

_____
_____
_____
_____
_____
_____
_____
_____
_____

You should expect to eventually write common assessments that would score fours in all eight categories of the assessment evaluation. For deeper clarification and explanation on each of the evaluation criteria in the assessment tool, use figure 2.2 to reflect on how your team rates and scores on your common assessment instruments.

TEAM RECOMMENDATION

Design High-Quality Common Unit Assessments

Use figures 2.1 and 2.2 to determine the current strengths and areas to improve on your team common assessments.

- Commit to creating your common unit assessments before the unit begins.
- Respond to the prompt, What is your team plan to improve your common unit assessments this year?

The first four design criteria of the assessment instrument evaluation tool (identification of and emphasis on essential learning standards, balance of higher- and lower-level-cognitive-demand tasks, variety of assessment-task formats, and appropriate scoring agreements) are the most important aspects of many mathematics unit assessments—both during and at the end of a unit. These elements of assessment design are vital to how your assessments will work (see the Steve Jobs epigraph for this chapter on page 17) and how the assessments will be used by your team *and* by your students.

The first four mathematics assessment design criteria can often create places of *great inequity* in your mathematics assessment process and professional work. For that reason, team analysis of and reflection on the following explanations for the first four criteria will help your and your colleagues' growth and development in designing your common mathematics assessments.

Identification of and Emphasis on Essential Learning Standards

Both the common assessment formative process rubric (figure 1.2, page 12) and the assessment instrument quality evaluation rubric (figure 2.1, page 18) start with your teacher team determining clear essential learning standards. In order to provide targeted and specific feedback to your collaborative team about instructional practices and student learning as well as meaningful feedback to students about their learning, the feedback must be tied to essential learning standards.

TEACHER *Reflection*

Review the criteria from figures 2.1 and 2.2. *How should you organize your assessments?* This is a good first question for your team to ask. Describe your current practice.

_____
_____
_____
_____
_____
_____

Currently, your team might design and organize your common assessments in many ways, such as by format (multiple choice first and constructed response second), difficulty (from easier tasks to more difficult tasks), or order (tasks placed in the order they were identified for the assessment).

However, these methods of assessment design do not serve the intended purpose and use of the assessments. The organizational driver for your mathematics assessment design needs to be the *essential learning standards*. As Margaret Smith, Michael Steele, and Mary Lynn Raith (2017) point out, "Setting goals is the starting point for all decision making" (p. 195), meaning your students and your teacher team should set goals for learning based on the essential standards for the assessment.

Thus, your collaborative team creates a common mathematics assessment, establishes the essential learning standards for the unit, and then designs the mathematics tasks or questions you believe represent evidence of student learning and proficiency for each standard. There should be no more than two to five essential learning standards on any common end-of-unit mathematics assessment and perhaps one or two on a common unit quiz or mid-unit assessment.

There are many names for the types of standards your team will use, such as *power standards*, *priority standards*, *promise standards*, and so on. When your teacher team works within a PLC culture and for the purposes of this book, we suggest using the term *essential learning standards* for the unit. These are the *big idea* standards you expect your students to learn. The essential learning standards are what teams use when creating the following:

1. Targeted and actionable feedback to students for reflection, learning, and goal setting

2. Team-designed student intervention and extension activities

3. Instructional decisions during and after a unit (See *Mathematics Instruction and Tasks in a PLC at Work, Second Edition* in this series.)

Your collaborative team then creates proficiency maps that identify the essential learning standards with which students should be proficient by the end of each unit. A quick overview of your unit pacing guides can help your team make connections between the essential learning standards students should learn, which in turn informs quality assessment tasks and a planned team response to student demonstrations of learning when using the assessments. (Visit **go.SolutionTree.com/MathematicsatWork** for examples of K–5 unit-by-unit, grade-level proficiency maps. See also the *Mathematics Unit Planning in a PLC at Work* books; Schuhl, Kanold, Barnes, et al., 2021; Schuhl, Kanold, Deinhart, et al., 2021; Schuhl, Kanold, Deinhart, Larson, & Toncheff, 2020; Schuhl, Kanold, Kanold-McIntyre, et al., 2021.)

Rick DuFour and Bob Marzano (2011) ask you to think of the essential learning standards as part of your team's guaranteed and viable curriculum for each unit. A *guaranteed curriculum* is one that promises the community, staff, and students that a student in school will learn specific content and processes regardless of the teacher the student receives. A *viable curriculum* refers to ensuring there is adequate time for all students to learn the guaranteed curriculum during the school year.

Thus, clear team agreement on the essential learning standards and the rigor of those standards for the unit is a priority. It is the starting point for answering PLC critical question 1 (DuFour et al., 2016): What do we want all students to know and be able to do? This is the launching point for more equitable learning experiences for every student in your grade level or course.

The two to five essential learning standards also inform your team's intervention design and response to critical questions 3 and 4, discussed in chapter 6 of this book (page 121).

TEACHER *Reflection*

Look closely at your students' current assessments. How many essential learning standards are you listing on the assessments? Too many? Too few? If you do not list the standards, can you identify how many the tests assess?

_____
_____
_____
_____
_____
_____

Ultimately, the reason for designing your common assessments around the organizational driver of the essential learning standards is to identify where to target future emphasis for student learning, including:

1. Mathematical tasks or questions, such as the proper distribution and ensured standards alignment of those mathematical tasks placed on the common assessment
2. Team-designed student intervention and extension activities
3. Instructional decisions during and after a unit (We discuss this more thoroughly in *Mathematics Instruction and Tasks in a PLC at Work, Second Edition* in this series.)

Be sure your mathematics assessments, unit tests, and quizzes do not list too many essential learning standards. In order to design high-quality mathematics assessments your team and students can use for formative purposes, focus only on the big ideas and most essential learning standards. Also, be sure to write the essential learning standards in student-friendly language using *I can* statements. The *I can* statements help students verbalize and own their role in the assessment, learning, and reflection process designed to strengthen their efficacy, agency, and identity.

However, the essential standards used to create your assessments do not provide enough detail for your daily lesson learning targets. You most likely use or have heard of many terms for your lesson focus, such as *learning objective, learning target, daily objective, daily learning standard, lesson objective*, and so on. These more detailed and specific concepts and skills for the daily lessons help you and your students make sense of the essential learning standards (the bigger mathematics standards for the unit) and inform the types of tasks (questions) to include on the common assessment. For the purposes of our PLC work, we reference these specific parts of standards as *learning targets*.

Your daily learning targets help support the tasks selected for your common assessments but are too specific for test organization purposes. The essential learning standards in each unit comprise sections of the assessment and allow for student reflection and continued learning. (See pages 100–107 for specific student reflection examples.)

To help you compare and contrast essential learning standards and daily learning targets, figure 2.3 (page 24) provides a sample that illustrates a grade 4 fractions unit. The formal unit standards in the left column are the state or provincial standards from which the essential learning standards are generated.

The middle column shows essential learning standards as a teacher team would write them on the assessment or test in student-friendly *I can* language. The right column unwraps each standard into daily learning targets for your lesson design and planning using combinations of verbs and noun phrases in the original formal standard. (Visit **go.SolutionTree.com/MathematicsatWork** or scan the QR code that appears at the end of this chapter to find an unwrapping standards protocol from the *Mathematics Unit Planning in a PLC at Work* books.) The daily learning targets aligned to each essential learning standard define the success criteria needed for students to be proficient with each essential learning standard.

Using the sample grade 4 fractions unit, your team could create shorter common assessments for each of the essential learning standards throughout the unit. You and your students could then use the results to proactively receive feedback for additional instruction and help before the common end-of-unit assessment. (Visit **go.SolutionTree.com/MathematicsatWork** or scan the QR code to find similar models for first grade, seventh grade, and high school.)

TEACHER *Reflection*

How does your team create and develop its understanding of the essential learning standards for the unit?

How do you break down the essential learning standards for assessment purposes into daily learning targets for instruction purposes?

_____
_____
_____
_____
_____
_____
_____

Formal Unit Standards (Generic state or provincial standard language)	Essential Learning Standards for Assessment and Reflection (Student-friendly language)	Daily Learning Targets (What students should know and be able to do; unwrapped standards)
1. Explain why a fraction $\frac{a}{b}$ is equivalent to a fraction $\frac{(n \times a)}{(n \times b)}$ by using visual fraction models, with attention to how the number and size of the parts differ, even though the two fractions themselves are the same size. Use this principle to recognize and generate equivalent fractions.	I can explain why fractions are equivalent and create equivalent fractions.	• Explain why a fraction $\frac{a}{b}$ is equivalent to a fraction $\frac{(n \times a)}{(n \times b)}$ by: ▸ Using visual fraction models ▸ Telling how the number and size of the parts differ even though the number and size of the parts are the same size • Generate equivalent fractions.
2. Compare two fractions with different numerators and different denominators, for instance, by creating common denominators or numerators or by comparing to a benchmark fraction such as $\frac{1}{2}$. Recognize that comparisons are valid only when the two fractions refer to the same whole. Record the results of comparisons with symbols >, =, or <, and justify the conclusions.	I can compare two fractions and explain my thinking.	• Compare two fractions with different numerators and different denominators by: ▸ Creating common numerators ▸ Creating common denominators ▸ Comparing to a benchmark fraction • Record fraction comparisons using <, =, or >. • Justify fraction comparisons (for instance, using visual models). • Recognize that comparisons are valid only when the two fractions refer to the same whole.
3. Understand a fraction $\frac{a}{b}$ with $a > 1$ as a sum of fractions $\frac{1}{b}$. a. Understand addition and subtraction of fractions as joining and separating parts referring to the same whole. b. Decompose a fraction into a sum of fractions with the same denominator in more than one way, recording each decomposition by an equation. Justify decompositions, for instance, by using a visual fraction model. c. Add and subtract mixed numbers with like denominators (for example, by replacing each mixed number with an equivalent fraction, and/or by using properties of operations and the relationship between addition and subtraction).	I can add and subtract fractions and show my thinking.	• Compose a fraction using unit fractions. • Add fractions by joining parts of the same whole. • Subtract fractions by separating parts of the same whole. • Decompose a fraction more than one way into a sum of fractions with like denominators. • Add mixed numbers with like denominators using: ▸ Equivalent fractions ▸ Properties of operations ▸ The relationship between addition and subtraction • Subtract mixed numbers with like denominators using: ▸ Equivalent fractions ▸ Properties of operations ▸ The relationship between addition and subtraction

Quality Common Mathematics Assessments

Formal Unit Standards (Generic state or provincial standard language)	Essential Learning Standards for Assessment and Reflection (Student-friendly language)	Daily Learning Targets (What students should know and be able to do; unwrapped standards)
4. Apply and extend previous understandings of multiplication to multiply a fraction by a whole number. a. Understand a fraction $\frac{a}{b}$ as a multiple of $\frac{1}{b}$. For example, use a visual fraction model to represent $\frac{5}{4}$ as the product $5 \times (\frac{1}{4})$, recording the conclusion by the equation $\frac{5}{4} = 5 \times (\frac{1}{4})$. b. Understand a multiple of $\frac{a}{b}$ as a multiple of $\frac{1}{b}$, and use this understanding to multiply a fraction by a whole number. For example, use a visual fraction model to express $3 \times (\frac{2}{5})$ as $6 \times (\frac{1}{5})$, recognizing this product as $\frac{6}{5}$. (In general, $n \times (\frac{a}{b}) = \frac{(n \times a)}{b}$.)	I can multiply a fraction by a whole number and explain my thinking.	• Multiply a fraction by a whole number. • Show that a fraction $\frac{a}{b}$ is a multiple of $\frac{1}{b}$ and write an equation to match. • Multiply a fraction by a whole number using the idea that $\frac{a}{b}$ is a multiple of $\frac{1}{b}$. ‣ Use a visual model. ‣ Write expressions.
3d. Solve word problems involving addition and subtraction of fractions referring to the same whole and having like denominators (for example, by using visual fraction models and equations to represent the problem). 4c. Solve word problems involving multiplication of a fraction by a whole number (for example, by using visual fraction models and equations to represent the problem). For example, if each person at a party will eat $\frac{3}{8}$ of a pound of roast beef, and there will be 5 people at the party, how many pounds of roast beef will the party need? Between what two whole numbers does your answer lie?	I can solve word problems involving fractions.	• Solve word problems involving addition and subtraction of fractions with like denominators referring to the same whole using: ‣ Visual fraction models ‣ Equations • Solve word problems involving multiplication of a fraction by a whole number. ‣ Use a visual model. ‣ Write equations.

Figure 2.3: Sample grade 4 fractions unit—Essential learning standards.

Visit go.SolutionTree.com/MathematicsatWork for a free reproducible version of this figure.

When you work with your colleagues to determine the essential learning standards for common unit assessments, you ensure equitable learning expectations and experiences for students in classrooms across your team.

> **TEAM RECOMMENDATION**
>
> ## Determine Essential Learning Standards for Student Assessment and Reflection
>
> - Identify the formal learning standards students must be proficient with by the end of the unit (state or provincial standards).
> - As a teacher team, make sense of the formal learning standards and discuss what students must know and be able to do to demonstrate proficiency (daily learning targets).
> - Group the formal standards as needed to create two to five essential learning standards for the unit. Write these using student-friendly language and begin each with *I can* statements.

Once your team begins the common assessment design (your high-quality surfboard) using the essential learning standards for the unit, what's next? The next step is revealed in your choices for the mathematical tasks you believe align to and represent each essential learning standard on the assessment. After you determine the essential learning standard, your choice of tasks drives everything you do because "tasks provide the vehicle that move your students from their current understanding toward complete understanding of the essential learning standard" (Smith et al., 2017).

Choosing mathematical tasks is one of your most important professional responsibilities as a mathematics teacher. It permeates everything you do. You decide each day the mathematical tasks with which students will engage in class. You then decide the tasks and questions to assign for homework. You also work with your collaborative team to decide the nature of the mathematical tasks you place on all common assessments, as well as the tasks used for intervention on essential standards. The tasks chosen determine how deeply students make sense of the mathematics they are learning and also provide an opportunity to be inclusive of the student cultures represented in your classroom.

That is a lot of power.

And, it is why working on these decisions *together* as a team is so important. It is why the second criterion in figure 2.1 (page 18) is the rigor revealed by the cognitive demand of the mathematical tasks you have chosen for the assessment.

Balance of Higher- and Lower-Level-Cognitive-Demand Mathematical Tasks

There are several ways to label the cognitive demand or rigor of a mathematical task; however, for the purposes of this book, mathematical tasks are classified as either lower-level cognitive demand or higher-level cognitive demand, as Margaret S. Smith and Mary Kay Stein (1998) first defined in their task analysis guide printed in full in the appendix: "Cognitive-Demand-Level Task Analysis Guide" (page 137).

Lower-level-cognitive-demand tasks typically focus on memorization by performing standard or rote procedures without attention to the properties that support those procedures (Smith et al., 2017; Smith & Stein, 2011). *Higher-level-cognitive-demand tasks* are those for which students do not have a set of predetermined procedures to follow toward a solution, or if the tasks involve procedures, they require that students justify why and how to perform the procedures. Additionally, when selecting tasks, your teacher team has an opportunity to honor your students' cultures and interests.

Figure 2.4 shows examples of lower- and higher-level-cognitive-demand tasks (for use either in class or on an assessment) for various grade levels. Use the most appropriate grade-level question, and discuss why each question meets its cognitive-demand level using the descriptions that the "Cognitive-Demand-Level Task Analysis Guide" provides. The third-grade task gives an example of using a culturally relevant task as a way for students to see themselves in the mathematics they are learning (National Council of Supervisors of Mathematics [NCSM], 2022).

Quality Common Mathematics Assessments

Directions: Choose the most appropriate grade level that follows and discuss why each of the questions meets the cognitive-demand level to which it's assigned using the descriptions for lower- and higher-level-cognitive-demand tasks in the appendix (page 137).

Kindergarten: I can count up to 20 objects and write how many I counted.

Lower-level-cognitive-demand task:	Higher-level-cognitive-demand task:
Give students a worksheet with pictures of up to 20 objects to count. Students count the objects on the worksheet and next to each object write the numeral that represents the object.	Place various collections of objects in plastic bags around the classroom (collections should vary from 7 to 20 objects). Students move around the room to count the collection at each station and record the numeral on their recording sheet. The recording sheet has a picture of an object from each collection so students know where to write the numeral. As the students are counting, take observational notes on how students count their collection. Examples include the following. • Do students place the objects in a line, group the objects, or have them in a scattered formation? • Are students able to get the correct final count? • Are students touching the objects as the count? • Are there any numbers students skip when they count? • Are some quantities more difficult for students to count? If so, why? • Ask students how they are counting their collections.

Grade 3: I can find the product of two numbers.

Lower-level-cognitive-demand task:	Higher-level-cognitive-demand task:
Paola buys 6 boxes of brigadeiros (a Brazilian dessert). Each box holds 10 brigadeiros. How many brigadeiros does Paola have in all?	Paola is going to buy 100 Brazilian desserts for her family party. She needs to buy 4 different types of desserts and she can only buy whole boxes. Here are the different types of desserts the bakery sells. 	Dessert	Number of Portions of the Dessert in Each Box	 |---|---| 	Brigadeiro	10 in each box	 	Pé-de-moleque	5 in each box	 	Cocada	4 in each box	 	Branquinho	6 in each box	 How many boxes of each type of dessert can Paola buy to have 100 desserts? Show or explain your thinking.

Figure 2.4: Examples of lower- and higher-level-cognitive-demand tasks.

continued →

Grade 6: I can use ratio and rate to solve real-world problems and mathematical problems.

Lower-level-cognitive-demand task:	**Higher-level-cognitive-demand task:**
Four single-scoop ice cream cones cost $14.20. At that rate, how much do 3 single-scoop ice cream cones cost?	Natasha has a bag that contains 225 coins, either nickels or pennies. There are 4 nickels for every 11 pennies in the bag. • How many nickels and how many pennies are in the bag? Explain how you determined your answer. • What is the total value of the coins in the bag?

High school: I can solve a system of linear equations using algebra.

Lower-level-cognitive-demand task:	**Higher-level-cognitive-demand task:**
Solve the system of equations using the substitution or elimination method. $$\begin{cases} 2x + 3y = 51 \\ y = -2x + 37 \end{cases}$$	Write and then solve a system of linear equations to determine the unit cost of a baseball, a hat, and a bat. |	Couple's Package	Autograph Hunter Package	Let's Play Ball Package
---	---	---	---
	$46.00	$31.00	$33.00	

Source: Adapted from Howard County Public School System Office of Secondary Mathematics, 2022. Used with permission.

Visit go.SolutionTree.com/MathematicsatWork for a free reproducible version of this figure.

TEACHER *Reflection*

After using figure 2.4 and the appendix (page 137) to explain why the grade-level tasks are low level or high level, think about a concept that students in your grade level must learn, and write or describe a lower-level-cognitive-demand task and a higher-level-cognitive-demand task for that concept.

Describe how your cognitive-demand choices for each task reveal underlying student understanding of an essential learning standard.

_____
_____
_____
_____
_____
_____
_____
_____
_____
_____

A key word regarding mathematical rigor is *balance*, as noted in criterion 3 in figure 2.2 (page 19). Your common assessment tasks should reveal a balance of procedural fluency tasks and conceptual understanding tasks with applications of each type. Proficiency on all common mid- and end-of-unit assessments should be based on both types of tasks. Generally, the rigor-balance ratio should be about 60/40 (lower- to higher-level cognitive demand) on an end-of-unit assessment for essential standards requiring higher-level cognitive demand for student proficiency. When planning the unit, your teacher team makes sense of the essential learning standards to determine which ones require learning at a higher level (see the *Mathematics Unit Planning in a PLC at Work* series).

TEACHER *Reflection*

Think about your current end-of-unit assessments. In general, describe the balance of lower-level-cognitive-demand questions to higher-level-cognitive-demand questions on your assessments.

_____
_____
_____
_____
_____
_____
_____
_____
_____
_____
_____
_____
_____
_____
_____
_____
_____
_____
_____

Your collaborative team should not try to create every mathematics assessment task used on a common unit assessment. Your team may decide to start with a publisher assessment or previously used assessment to gather quality problems, questions, and tasks. You may also search websites for assessment questions and tasks using some of the online resources for mathematics assessment support (visit **go.SolutionTree.com /MathematicsatWork** or scan the QR code at the end of the chapter for the free reproducible "Online Resources Reference Guide for Mathematics Support").

You can access tasks in textbooks or other reliable resources. Your teacher team may create some of your own assessment questions to gather evidence of student learning as you build assessments together. Finally, your team should discuss examples of lower- and higher-level-cognitive-demand tasks you write when planning a unit (see the *Mathematics Unit Planning in a PLC at Work* series).

If you are a singleton teacher, and you are teaching a grade level or course with released state or provincial assessment questions or released national assessment questions, consider placing some of these tasks on your assessments to compare the responses of your students to those of comparable students in your state or province or across the country. Additionally, if you are a singleton teacher on a vertical collaborative team, consider how you will assess a common skill using content in each grade or course represented on the team (skills may include process standards such as critiquing the work of others, or content skills such as drawing pictures or graphs or solving equations with shown work). Determine the tasks to use on each grade-level or course-based assessment, and share assessments for feedback across the team. Another option if you are a singleton teacher is to find a course-alike colleague in a different school and work together virtually or in person utilizing a digital team folder to share common assessments and other team artifacts.

When you work with colleagues to agree on the lower- and higher-level-cognitive-demand task balance for your assessments, you create improved rigor expectations for student learning and erase the inequities caused by individual interpretation of rigor from teacher to teacher on your team.

No matter where you source your assessment questions and tasks, you most likely will encounter many different formats of questions. The third criterion for you and your team to consider, then, as you build a high-quality assessment is variety of assessment-task formats.

These choices for assessment format, when made in isolation from your colleagues, can also become a source of inequity in your assessment of students if you do not agree as a collaborative team. Discussion of these formatting choices is next.

TEAM RECOMMENDATION

Agree On the Lower- and Higher-Level-Cognitive-Demand Mathematical Tasks Chosen for Common Unit Assessments

- As a team, identify examples of lower- and higher-level-cognitive-demand tasks for the essential learning standards in a unit.

- Determine which tasks to place on the assessment, and use a 60/40 ratio of low- to high-level-cognitive-demand tasks as a guide for ensuring a balance of rigor on the assessment (standards in preK and kindergarten may not always require higher-level-cognitive-demand tasks).

Variety of Assessment-Task Formats

Assessment formats reflect how you choose to present the tasks on each assessment. When you are designing most common mid-unit or end-of-unit assessments, your mathematics assessment tasks will include either selected-response tasks or constructed-response tasks.

- *Selected-response tasks* refer to multiple choice (one or more correct answers), fill-in-the-blank, matching, true or false, or questions with one or more immediate right or wrong answers.

- *Constructed-response tasks* refer to a short answer in which students show their reasoning or solution pathway as well as their final answer. In preK, kindergarten, and early first grade, this would also include a short answer in oral or demonstration form instead of a written expectation. Your team should determine if you will administer the mathematics assessment one-on-one, in small groups, or with a whole group. Written performance assessment tasks or projects are in the family of constructed-response tasks, as students are showing learning, in real time, which is often complex in reasoning.

> **TEACHER** *Reflection*
>
> Currently, what is your favorite format or problem type to use for the mathematics problems or tasks on your common assessments? Do you prefer multiple choice, open ended, true or false, or matching? Do you prefer to give the assessments online or use paper and pencils? Why do you have these format preferences?
>
> _____
> _____
> _____
> _____
> _____
> _____
> _____
> _____
> _____
> _____

Ideally, your common assessments provide students with practice using these various assessment formats. And, whether you determine the assessment is given online or on paper, students will be required to show their reasoning and work in order to receive meaningful feedback about their demonstrations of learning, and to receive at least partial credit for their solution pathways. They may write their solutions on paper, upload pictures of their paper solutions, or use a touch screen to show their reasoning, to name a few options.

Selected Response

Often, the most difficult selected-response tasks to design and use accurately are multiple choice and matching. Guidelines are provided to help you evaluate the quality of these two formats.

Multiple Choice

Consider the following five guidelines when using multiple-choice tasks.

1. Be sure every distractor (incorrect answer) reflects a common misconception or error students often demonstrate for the concept.

2. Be sure each incorrect answer provides your teacher team with as much information about student learning as a correct response would. If your team cannot think of plausible distractors, consider leaving the question open ended this year and gathering common misconceptions as distractors for next year's assessment.

3. Consider allowing students to identify more than one correct answer, which more easily allows for higher-level-cognitive-demand tasks. Be sure to let students know in the directions or the stem (question) that more than one answer may be correct.

4. Consider parallel construction in the responses. If, for example, one distractor or answer is written as a decimal while the rest are written as whole numbers, students often choose the one that looks different from the rest, whether correct or incorrect. Include at least two decimals in this case. The answer and distractors should be similar in length whether written as sentences or number values.

5. Be sure to allow for partial credit (students can show work) when it is time to score multiple-choice assessment questions.

Matching

Consider the following two guidelines when using matching tasks.

1. Make the list students match from shorter than the list they match to. This prevents faulty data. When the lists are the same length, many students simply connect the last remaining parts when they get to the last item. If they have made an error previously, they now have made at least two errors in reasoning. If the list they are matching to is longer, students are more apt to look at the remaining options and correct their thinking as needed.

2. Allow students to match to any possible response more than one time when matching, again to minimize flawed results from the assessment.

Constructed Response

Consider the following four guidelines when using constructed-response tasks.

1. Be sure the directions for the task explain the work you require of the students. For example, if you and your colleagues expect students to draw a picture or graph and that is part of your scoring agreement, you must tell the students that they need to include a picture or graph in their solution.

2. Keep the task's wording simple except for academic vocabulary the task requires.

3. Write tasks using familiar contexts for students. For example, avoid a skiing task if students live in an area without easy access to skiing.

4. Be careful when chunking a constructed-response item into several parts (steps) students complete to answer the overall question. Allow students to show their thinking through their choice of strategies or solution pathways without reducing the cognitive demand of the task by breaking it down into too many parts.

TEACHER *Reflection*

Which of the guidelines for selected-response tasks of multiple choice and matching or constructed-response tasks do you already consider when creating your unit assessments?

Which guidelines will your team need to remember when choosing or creating assessment tasks to make your common unit assessments stronger?

_____
_____
_____
_____
_____
_____
_____
_____
_____
_____

TEAM RECOMMENDATION

Make a Plan for Assessment Methods

- Determine which assessment tasks should be selected response and constructed response.

- Examine the guidelines for writing quality selected-response and constructed-response tasks.

- Reach agreement on how students will share their reasoning when solving tasks presented in a multiple-choice or selected-response format whether the assessment is given online or on paper.

Once the assessment is constructed, you now work as a team on the fourth criterion for writing high-quality common assessments: appropriate and clear scoring agreements.

Appropriate and Clear Scoring Agreements

At this point, your collaborative team has created a common mathematics assessment based on learning targets and mathematical tasks that align to the essential learning standards for the unit. Your team has also determined the best cognitively balanced mathematical tasks needed to gather evidence of student learning. How, then, should your team grade, score, and evaluate each assessment task to provide meaningful feedback to the students, you, and your team members? Reaching agreement on this issue is a primary responsibility of your professional work together.

Scoring an assessment task using points is a two-step process.

1. The teacher team reaches agreement on the scoring value of each assessment task.

2. The teacher team reaches agreement on the evidence-of-understanding requirements for a student to receive full or partial credit for a response to each mathematical task.

When scoring an assessment using a proficiency rubric, teams clarify what evidence of learning students must show to be proficient with the selected tasks and then determine overall for each essential learning standard the level of proficiency shown by the student.

TEACHER *Reflection*

Think about your current assessments.

If you score assessments using a points-per-question scale, do you clearly state on the test the points possible for each task or mathematics problem? Does your total points for the task make sense based on the task's complexity of reasoning required for successful student demonstration of work?

If you use a standards-based proficiency rubric, does each team member agree on the meaning of 1, 2, 3, or 4 as a grade for each essential learning standard on the assessment? How do you determine a student's level of proficiency for each essential learning standard using the group of questions associated with each essential learning standard?

_____
_____
_____
_____
_____
_____
_____
_____
_____
_____

Review the assessment questions in figure 2.5 (page 34) and select a grade level most appropriate to your current teaching assignment. First, determine how many points you would use to score the task (scoring guide) or how you would describe proficiency levels 1–4 for the task (proficiency rubric). If you are using a proficiency rubric, 1 represents minimal understanding, 2 represents partial understanding, 3 represents proficient (or adequate) understanding, and 4 represents exceeding grade-level or course-based understanding or expectations.

Then, as a collaborative team, use the team discussion tool in figure 2.6 (page 35) to calibrate your scoring decisions with your colleagues. (Visit **go.SolutionTree.com/MathematicsatWork** or scan the QR code at the end of the chapter to find additional grade-level tasks online.)

How did you and your teammates determine the scoring of the mathematical task for your selected grade-level problem? Do students receive one point for their reasoning and one point for the correct answer, or no credit at all if the answer is incorrect? Is each part of the question worth one point, two points, four points, or more? What must a student show to earn a score of 1, 2, 3, or 4 if you used a proficiency rubric? If you give an assessment in real time orally, how do you document the score and account for student reasoning?

TEACHER *Reflection*

How similar or different were your and your team members' ideas for scoring the task you chose in figure 2.5 (page 34)?

Why is it important to reach common scoring agreements for the tasks on your assessments?

_____
_____
_____
_____
_____
_____

Directions: Choose a grade-level task most appropriate to your current teaching assignment. Determine how many points you would use to score the task or describe how you would give a proficiency-based score of 1–4 for the task.

Kindergarten Task

Essential learning standard: I can solve addition and subtraction word problems within 10 and show my thinking.

Sam has 8 toy cars. He gives 3 toy cars to his brother. How many toy cars does Sam have now? Show how you know your answer is correct.

Grade 3 Task

Essential learning standard: I can represent and solve two-step word problems using the four operations.

Jo buys 5 small baskets of strawberries. Each basket holds 9 fresh strawberries. She uses 15 strawberries for a dessert. How many strawberries does Jo have left over? Use words, numbers, pictures, or a combination of these to show how you know your answer is correct.

Grade 6 Task

Essential learning standard: I can solve word problems by writing equations and solving them. (Equations are of the form $x + p = q$ and $px = q$ for cases in which p, q, and x are all nonnegative rational numbers.)

Joshua makes a dessert. He uses 5 equal scoops of sugar and uses a total of $3\frac{1}{2}$ cups of sugar. How much sugar is in each scoop? Use words, numbers, pictures, or a combination of these to show how you know your answer is correct.

High School Task

Essential learning standard: I can graph quadratic functions and analyze intercepts, maxima, and minima in the context of real-world situations.

At the annual Halloween Pumpkin Chunkin' event, the winning team launched a pumpkin from a trebuchet.* The height (h) of the pumpkin, in feet, can be modeled by the function $h(x) = -16t^2 + 120t + 34$, where t represents the time, in seconds, the pumpkin was in the air after it was launched. Use graphing technology to respond to the following.

  a. What was the height of the pumpkin when it was launched? Explain how you determined your answer.
  b. What was the maximum height of the pumpkin during its flight? How long was the pumpkin in the air before it reached its maximum height?
  c. How long was the pumpkin in the air before it hit the ground? Describe how you could determine the answer using the function.

*A trebuchet is a medieval war device that was used to sling large boulders into castle walls or opposing troops.

Figure 2.5: Teacher team discussion tool—Sample assessment questions for team scoring.

Visit go.SolutionTree.com/MathematicsatWork for a free reproducible version of this figure.

Collaborative-Team-Task-Scoring Discussion Prompts

Directions: With your collaborative team, examine the following questions for the task you chose from figure 2.5.

1. How would you assign the points for different parts of the solution or describe levels of proficiency using a scale of 1–4?

2. If the point value for the task is greater than 1, how can a student earn partial credit? If you use a proficiency rubric, how might a student earn a 2?

3. What evidence of student learning must be shown to receive full credit or demonstrate proficiency?

4. How would your team ensure the scoring of the task is consistent among all team members?

5. How is the expected scoring (points or proficiency rubric) for the assessment task or tasks revealed to the students?

Figure 2.6: Teacher team discussion tool—Scoring assessment tasks.

*Visit **go.SolutionTree.com/MathematicsatWork** for a free reproducible version of this figure.*

If your collaborative team does not engage in conversations about scoring, inequities will persist across your team, and ultimately, assigned grades and feedback will be inaccurate from teacher to teacher. Your team needs to articulate and agree on the scoring guide (based on the complexity of reasoning appropriate to the mathematical tasks) or proficiency rubric for each essential learning standard before students take the test.

Figure 2.7 shows possible scoring agreements for the tasks in figure 2.5 (page 34), accounting for complexity of reasoning as well as the nature of the students' conclusions.

Earning a 4 on a proficiency rubric is difficult for a single task and is more accurately applied to an essential learning standard. The 4 offers ideas for student work that exceed grade-level expectations for proficiency. (You can find more information about proficiency rubrics on page 70.)

Having strong norms in place will help your collaborative team learn how to agree on the scoring of your common assessments. Initially, you might need a facilitator, whether an instructional coach or other teacher or support person. This person can help you navigate the potentially difficult conversations when you have a wide variance of opinions for how to evaluate, score, and give feedback on student work.

Figure 2.8 (page 39) provides sample student work for each task from figure 2.5 (page 34). Using your team scoring agreements for the task, evaluate and score student work for the task appropriate to your grade level. Consider how your team would score student work on the assessment question based on the complexity of reasoning required by the student. Does the student work demonstrate knowledge at a proficiency level or greater? How would you know? Once again, you can use the teacher team discussion prompts in figure 2.6 (page 38) to help team conversations around common scoring, using the selected assessment task. It might be interesting to "blind" double score the task with a partner and discover how similar your scores are for the same student work.

If you are grading the kindergarten task, how do the scores for the two students differ, if at all, knowing student A has the incorrect answer and student B uses the incorrect symbol to show their thinking? If you are grading the third-grade task, how might counting errors influence the student's overall score? If you are grading the sixth-grade task, how might the check that verified the incorrect answer factor into the overall score, if at all? If you are grading the high school task, how do the rounded answers influence the overall score? Is function notation or an equals sign needed in the work for part a?

TEACHER *Reflection*

If your team scores using a scoring guide (points), how can students earn partial credit for their work?

If your team scores using a proficiency rubric, what types of mistakes can a student make and still be considered proficient? How will your team determine proficiency (level 3)?

_____
_____
_____
_____
_____
_____
_____
_____
_____
_____
_____
_____

When you score common mathematics assessments privately and, therefore, often differently from one another, inequities can result in feedback and grading reports to students and team member analysis of results. To gather student work to score, determine how to gather student thinking and reasoning on the assessment. If students take the assessment on a computer, will they show their work on paper or upload their work for actionable feedback and continued learning?

Directions: Analyze the scoring guide points or proficiency rubric scores assigned to each task. Do you and your colleagues agree or disagree? Why?

Kindergarten Task

Essential learning standard: I can solve addition and subtraction word problems within 10 and show my thinking.

Sam has 8 toy cars. He gives 3 toy cars to his brother. How many toy cars does Sam have now? Show how you know your answer is correct.

Scoring Guide	Proficiency Rubric
Two points total: • 1 point for equation or work showing $8 - 3 = ?$ or $3 + ? = 8$ • 1 point for answer (5 toy cars) with or without units	4—Student solves the task correctly with work to show the answer is correct and checks the work or shows two ways to justify the answer, including the units (toys, cars, or toy cars) at the end of the answer. 3—Student solves the task correctly with work to support the answer and may have reversal of numbers or not have the units as part of the answer. 2—Student has correct work with an incorrect answer or a correct answer, but the work does not support the answer. 1—Student shows minimal understanding in the work and the answer. (Student might solve $8 + 3 = ?$ correctly.)

Grade 3 Task

Essential learning standard: I can represent and solve two-step word problems using the four operations.

Jo buys 5 small baskets of strawberries. Each basket holds 9 fresh strawberries. She uses 15 strawberries for a dessert. How many strawberries does Jo have left over? Use words, numbers, pictures, or a combination of these to show how you know your answer is correct.

Scoring Guide	Proficiency Rubric
Three points total: • 1 point for finding the total number of strawberries ($5 \times 9 = 45$) • 1 point for finding the strawberries left (for example, $45 - 15 = 30$) • 1 point for answer (30 strawberries)	4—Student solves the task correctly with work to show the answer is correct and checks the work or shows two ways to justify the answer, including the units (strawberries) at the end of the answer. 3—Student solves the task correctly with work to support the answer and may not have the units as part of the answer. 2—Student has correct work with an incorrect answer or a correct answer, but the work does not support the answer. 1—Student shows minimal understanding in the work and the answer. (Student might use the numbers 5, 9, and 15 with random operations and perform the calculations correctly.)

Grade 6 Task

Essential learning standard: I can solve word problems by writing equations and of the form $x + p = q$ and $px = q$.

Joshua makes a dessert. He uses 5 equal scoops of sugar and uses a total of $3\frac{1}{2}$ cups of sugar. How much sugar is in each scoop? Use words, numbers, pictures, or a combination of these to show how you know your answer is correct.

Scoring Guide	Proficiency Rubric
Three points total: • 1 point for knowing how to solve $3\frac{1}{2} \div 5 = ?$ or $5 \times ? = 3\frac{1}{2}$ • 1 point for performing the calculation correctly or using models correctly • 1 point for answer ($\frac{7}{10}$ cup)	4—Student solves the task correctly with work to show the answer is correct and checks the work or shows two ways to justify the answer, including the units (cup, of a cup, or cups) at the end of the answer. 3—Student solves the task correctly with work to support the answer and may not have the units as part of the answer. 2—Student has an error in the solution pathway though the answer matches the student work, or student has a correct answer, but the work does not support the answer. 1—Student shows minimal understanding in the work and the answer. (Student might use another operation with the numbers $3\frac{1}{2}$ and 5.)

Figure 2.7: Example of task scoring guides or proficiency rubric scores for tasks in figure 2.5.

continued →

High School Task

Essential learning standard: I can graph quadratic functions and analyze intercepts, maxima, and minima in the context of real-world situations.

At the annual Halloween Pumpkin Chunkin' event, the winning team launched a pumpkin from a trebuchet.* The height (h) of the pumpkin, in feet, can be modeled by the function $h(x) = -16t^2 + 120t + 34$, where t represents the time, in seconds, the pumpkin was in the air after it was launched. Use graphing technology to respond to the following.

a. What was the height of the pumpkin when it was launched? Explain how you determined your answer.
b. What was the maximum height of the pumpkin during its flight? How long was the pumpkin in the air before it reached its maximum height?
c. How long was the pumpkin in the air before it hit the ground? Describe how you could determine the answer using the function.

*A trebuchet is a medieval war device that was used to sling large boulders into castle walls or opposing troops.

Scoring Guide	Proficiency Rubric
Six points total: a. 2 points—1 point for identifying the current starting height as 34 feet. Then, 1 point for an explanation describing that when $x = 0$, the resulting height is 34 feet, or that upon inspection of the function, written in standard form, the value of the constant (c) is 34 feet. b. 2 points—1 point for correctly identifying the maximum height as 259 feet. Then, 1 point for correctly identifying 3.75 seconds as the time at which the pumpkin reached its maximum height. c. 2 points—1 point for correctly identifying 7.773 seconds as the time of impact. Then, 1 point for describing or showing "substitute the value of 0 for $h(x)$ and then solve the equation for x" or "use graphing technology to identify the positive zero, or x-intercept." 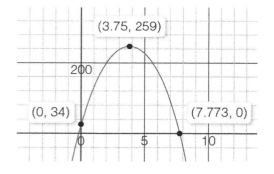	4—Student solves the task correctly and provides explanations that attend to precision and use academic language appropriately. The response indicates the student has a strong understanding of the connection between the function and the real-world scenario it is modeling and shows multiple representations as evidence. 3—Student solves the task correctly (process) but explanations do not fully attend to precision or they fail to use academic language. Or, the student makes a minor mistake that leads to an incorrect answer, but the student's explanations demonstrate that the student has a strong understanding of the connection between the function and the real-world scenario it is modeling. 2—Student reflects a conceptual error in the response. Student might assign correct numerical values but assign those values to incorrect phenomena (for example, 34 feet as the maximum height). Student's explanations are incomplete or inaccurate. 1—Student shows minimal understanding in the work shown and explanation. For example, the student may respond that 120, a value from the written function, is the maximum height or starting height. Explanations are completely inaccurate or missing.

Directions: Evaluate student work using your team scoring agreements. Is the student proficient? Does everyone on your team agree?

Kindergarten Task

Essential learning standard: I can solve addition and subtraction word problems within 10 and show my thinking.

Sam has 8 toy cars. He gives 3 toy cars to his brother. How many toy cars does Sam have now? Show how you know your answer is correct.

Student work:

Student A	Student B
8 − 3 = 6 (8 split into 3 and 6)	☒☒☒ ☐☐☐☐☐ 8 + 3 = 5

Grade 3 Task

Essential learning standard: I can represent and solve two-step word problems using the four operations.

Jo buys 5 small baskets of strawberries. Each basket holds 9 fresh strawberries. She uses 15 strawberries for a dessert. How many strawberries does Jo have left over? Use words, numbers, pictures, or a combination of these to show how you know your answer is correct.

Student work:

15 strawberries

9 + 6 = 15

(basket) + (basket) + (basket) + (basket) +

(basket) = 54

28 left over

Grade 6 Task

Essential learning standard: I can solve word problems by writing equations and solving them.
(Equations are of the form $x + p = q$ and $px = q$ for cases in which p, q, and x are all nonnegative rational numbers.)

Joshua makes a dessert. He uses 5 equal scoops of sugar and uses a total of $3\frac{1}{2}$ cups of sugar. How much sugar is in each scoop? Use words, numbers, pictures, or a combination of these to show how you know your answer is correct.

Student work:

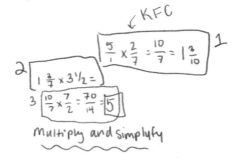

✓ KFC

1. $\frac{5}{1} \times \frac{2}{7} = \frac{10}{7} = 1\frac{3}{10}$

2. $1\frac{3}{7} \times 3\frac{1}{2} =$

3. $\frac{10}{7} \times \frac{7}{2} = \frac{70}{14} = 5$

Multiply and simplyfy

Figure 2.8: Sample assessment questions with student work for team scoring.   continued →

High School Task

Essential learning standard: I can graph quadratic functions and analyze intercepts, maxima, and minima in the context of real-world situations.

At the annual Halloween Pumpkin Chunkin' event, the winning team launched a pumpkin from a trebuchet.*
The height (h) of the pumpkin, in feet, can be modeled by the function $h(x) = -16t^2 + 120t + 34$, where t represents the time, in seconds, the pumpkin was in the air after it was launched. Use graphing technology to respond to the following.

  a. What was the height of the pumpkin when it was launched? Explain how you determined your answer.
  b. What was the maximum height of the pumpkin during its flight? How long was the pumpkin in the air before it reached its maximum height?
  c. How long was the pumpkin in the air before it hit the ground? Describe how you could determine the answer using the function.

*A trebuchet is a medieval war device that was used to sling large boulders into castle walls or opposing troops.

Student work:

a. $-16t^2 + 120t + 34$
   $-16(0)^2 + 120(0) + 34$
   $0 + 0 + 34$
   34 feet

b. The highest point was about (4, 260). That means it took 4 seconds for the pumpkin to reach its maximum hight of 260 feet.

c. I found the time of the pumpkin crash on my calculator. The x-intercept is about (8, 0) so it took 8 seconds for the pumpkin to hit the ground.

TEAM RECOMMENDATION

Determine Common Scoring Agreements for Common Mathematics Assessments

- Determine as a team how you will score assessment tasks on common assessments (points on a scoring guide or a proficiency rubric).
- Clarify how students earn points or how they earn a score of 1–4 using a proficiency rubric.
- Determine how students will know your scoring rubrics to be used on an assessment.

You have now explored the first four criteria for the assessment instrument quality evaluation rubric, which are:

1. Identification of and emphasis on essential learning standards (specific feedback to students)
2. Balance of higher- and lower-level-cognitive-demand mathematical tasks
3. Variety of assessment-task formats
4. Appropriate and clear scoring agreements

The remaining four criteria address overall formatting and clarity for the common assessment so student results are valid and reliable. Clear directions, academic language, visual presentation, and logistics (how the assessment is administered) are essential assessment design criteria that will impact students' accurate demonstrations of their mathematical understanding and competencies.

Clarity of Directions, Academic Language, Visual Presentation, and Logistics

Reaching team agreements on *how* to format assessment tasks with clear directions, language, and consideration of visual appearance, as well as clarifying the logistics for giving the assessment, will also influence student learning and results.

Planning for clarity of directions, academic language, visual presentation, and logistics means you and your colleagues can have confidence that the design structure of the assessment is not skewing results, and your students will most likely persevere when taking the mathematics assessment because it appears visually manageable and easy to read.

TEACHER *Reflection*

How do you currently decide, as a team, if the directions and vocabulary and symbol notations your assessment uses are appropriate for and clear to your students?

How do you know if students have enough space to show their work and if the assessment is administered consistently (including giving students enough time to finish)?

_____
_____
_____
_____
_____
_____

Clarity of Directions

In order for you and your collaborative team to know whether or not students are proficient with each essential learning standard, the directions need to clearly specify what students are expected to *do* to fully respond and answer each question. For example:

- For which mathematics tasks must students justify or explain their solutions?
- For which mathematics tasks must students make a table or draw a picture or graph?
- For which mathematics tasks can students use manipulatives or some type of technology?
- For which mathematics tasks must students write expressions or equations or build a model?

The directions clarify how students need to answer specific questions. Pay special attention to the verbs you use in the directions as they convey the action students must use to demonstrate understanding.

When you and your colleagues create an answer key for the common assessment and compare your solutions, you will quickly see if any two team members interpreted the directions differently. If different interpretations exist, work as a teacher team to decide whether both solutions are acceptable or if you need to clarify the directions further.

Sometimes you will collectively agree that the directions are clear only to find that students interpret them in a way you never anticipated. If this happens, discuss as a team what needs to be revised the next time you give the assessment and how you will score student work this time around.

Academic Language

The vocabulary and symbol notations in mathematics should be precise and comprise a language that allows students to make connections between mathematical concepts and clearly communicate their understanding of those concepts. When planning the unit, your teacher team identifies the mathematics vocabulary and notations students need to learn so they can communicate their reasoning throughout the unit (see the *Mathematics Unit Planning in a PLC at Work* series). The vocabulary and notations identified should match those on the assessment.

You and your team model precision when the tasks you place on the assessment use direct, fair, accessible, clearly understood language and consistent symbol notations. In return, you can expect students to demonstrate proficiency with concepts by also using precise academic language and notations in their solutions. You would not, for example, want a student to reference the *top* and *bottom* of a fraction but rather the *numerator* and *denominator*. Similarly, you would not want students to write *cos* or *sin* but rather *cos x* or *sin x*, and you would want them to understand the terms read as functions (*cosine of x* and *sine of x*). Precision in language and the solution process greatly influences effective mathematical communication.

Visual Presentation

When creating the assessment, consider how much space you allocate for students to explain their reasoning and illustrate their solutions. Is there enough? Too much? Are any graphs or pictures clearly visible and, if needed, can students complete a graph or picture based on its size? Is the font size large enough for students to easily read or make sense of the assessment items? Are any visuals used in preK or the primary grades large enough to make sense of (for example, dots to subitize, shapes, and numbers)?

Often publisher assessments do not leave enough space for students to work, and the font size may be too small. If a publisher assessment is used as is, students might cram their work into the margins or write an answer with no work shown, believing it is not necessary. Additionally, if tasks are not spaced appropriately or they contain a lot of words, students may find the assessment daunting and give up before ever stopping to read and persevere to solve the tasks.

Along with appropriate space for work, visual presentation refers to the exam's neatness and organization. Can students read the assessment and symbol notations clearly? Do you group assessment items by essential learning standard in a way that makes sense? Work together to make sure the assessment is visually inviting and clear.

Logistics

Logistics refers to how an assessment is given to students consistently across your teacher team. One consideration when creating your common assessment is the amount of time your students will have to take the assessment. Once you determine that, you then create the appropriate number of questions accordingly. Higher-level-cognitive-demand tasks will take more time. Time yourself as you complete full solutions to the assessment tasks. Then time how long it takes your students to complete the same assessment. Use these timings to "find your number." Your number is what you multiply your time by to better estimate how long it might take your students to complete the assessment. The ratio of the time it takes you to complete the assessment to the time it takes your students to complete the assessment may vary by course at the secondary level and by grade level in elementary school.

Compare estimates as a team to determine the duration of time you believe is appropriate for the assessment. How close are you to one another, and how close are you to the actual time it takes your students? Do your students receive equitable time expectations from all teachers on your team? If you decide to split the assessment into parts, clarify which parts students will take, in which order, and on which days. What are your teacher team agreements for if students do not finish the assessment in the time allocated?

If, however, you find the assessment is shorter or longer than you would like, your teacher team may need to revise the assessment to better match an effective time allotment. Shorter common mid-unit assessments should still leave time for instruction in a class period or block of time. Educators often plan common end-of-unit assessments for one class period or one block of mathematics time.

Time can also be an issue for kindergarten assessments or mathematics assessments conducted through one-on-one conversations or small-group sessions, such as at a kidney bean table. To minimize loss of class time, consider how to use written assessments for those essential learning standards from which you will get valid results (for example, writing numbers to twenty can be done as a whole group in kindergarten) and minimize the number of assessments needing one-on-one interactions.

Sometimes, you may need to decide as a collaborative team how to conduct one-on-one assessments for students unable to demonstrate their learning through writing, but this does not necessarily need to be the assessment method for all students due to the time structure it requires. For example, some preschoolers may be able to write and draw the mathematical reasoning while other students need to orally share their reasoning for a standard such as count to ten. Any one-on-one or small-group oral assessments require a written script for teachers to use during the assessment and agreements across the team to make sure the results come from equitable assessment experiences.

Another part of clarifying assessment logistics is to discuss such issues as whether students can have the assessment read to them, if anchor charts or desk name markers need to be removed, or if students can have access to notes or technological devices during the exam. Determine your agreed-on conditions for which students will take the common assessment to ensure results are as accurate as possible and student reasoning is not minimized.

Your teacher team may have students with predetermined plans for accommodations or modifications (for example, individualized education plans [IEPs], 504 plans, or English learner [EL] plans). Be sure any accommodations or modifications are consistently implemented across the team so students have the scaffolding supports they need to demonstrate grade-level or course-based learning.

Additionally, consider as a team how you will prepare students for a common assessment. Avoid giving a review that closely matches the assessment. Your teacher team will only know if students truly learned the essential learning standards by having students demonstrate their learning on novel tasks.

Finally, another important logistics question is whether an assessment will be given on paper or online and whether students can use manipulatives or technology during the common assessment.

Consider the following guidelines when using manipulatives or technology as part of the common mathematics assessment.

1. Agree as a teacher team on how you want to use technology as a tool. How will students be able to use the technology as a choice? How will the students be required to show all of their work, including when given a Scantron assessment?

2. Agree as a teacher team how you will score any work that requires the use of technology. How will students demonstrate their solution pathway based on the use of the technology, specifically when presented with multiple-choice tasks? How will students receive partial credit, including when using online adaptive assessments (with scaffolding)?

3. Build conceptual knowledge using student opportunities to interact with mathematics through digital or concrete tools on the assessment. Can you provide access to technology or manipulatives without lowering the cognitive demand or creating a crutch during the assessment for learning?

4. It is best to use manipulatives or technology tools on tasks you design to assess conceptual understanding, application, or modeling of mathematics.

Since middle school students can often use embedded technology on state or provincial assessments as early as sixth grade, and high school students taking the SAT, the ACT, or Advanced Placement (AP) exams also use calculators on all or portions of the assessment, it is wise to address the use of technology as part of your local common assessment tools as well. When students use technology tools and manipulatives during an assessment, they can learn through exploration and streamline computations focusing on more complex mathematical skills.

If you plan to use manipulatives or technology in your common assessment instruments, all the team members should also allow students to benefit from using the same manipulatives or specific technology as part of the student instruction and assessment experience. In every classroom, every student must have access to the agreed-on manipulatives or technology.

For example, if you are assessing sorting and classifying skills in kindergarten, be sure each team member uses the same manipulatives (colored bears or pattern blocks) to create equitable and meaningful assessment results for all students. Similarly, if you are using fraction tiles in the intermediate grades, ensure all students have access to or are told to use the same fraction tiles. At the secondary level, access to and the required use of graphing calculators or algebra tiles should be consistent across all classrooms and for all students.

The task directions, academic language, space to write, and logistics influence student feedback results. When directions are unclear, or tasks omit important vocabulary or notations, or questions are crammed together with limited space for student work, students will not receive enough time to complete the assessment thoroughly. Make sure your teacher team agrees on each of these four elements of effective high-quality assessment design.

— **TEACHER** *Reflection* —

Examine a recent common unit assessment.

Read the directions. How clear are they, and how well do they match the common scoring agreements? What verbs do the directions use?

Read the tasks. Which academic language (vocabulary and notations) do they include? What might you need to revise or add to make questions on the assessment clearer for students?

Look at your assessment. Is there space for student work? How crammed are the words on a page? Visually, does the assessment look doable?

Consider logistics. How many minutes did it take you to make a complete answer key? How many minutes did it take your students to complete the assessment? What is the teacher-to-student ratio you use when estimating the amount of time that it will take students to complete the assessment? What agreements did your team make about how to give the assessment? To which manipulatives, technology, or other tools do students need access to demonstrate proficiency on your common unit assessments? How will your team provide equal access to these tools as part of mathematics instruction and assessment?

_____
_____
_____
_____
_____
_____
_____
_____
_____
_____

TEAM RECOMMENDATION

Determine Clarity of Directions, Academic Language, Visual Presentation, and Logistics

Take the assessment as if you were a student, and then check your responses as a team to ensure the following.

- Directions are clear.
- Academic language (vocabulary and notations) is fair and appropriate.
- Space between tasks on a paper or digital assessment is appropriate for the amount of work a student needs to show.
- Logistics are clarified. Time allocated for the assessment is doable for students across the classrooms of your team. Thinking is not reduced based on the information on the walls or desks. Decisions are made about when students might choose to use technology or tools and how to ensure equal access to these tools for all students.

The design of your teacher team's common assessments impacts how the assessments can be formatively used. Figure 2.1 (page 18), the assessment instrument quality evaluation rubric, shares eight criteria to consider when your collaborative team designs common assessments for the purpose of providing your team and students with formative feedback.

1. Identification of and emphasis on essential learning standards (student-friendly language)
2. Balance of higher- and lower-level-cognitive-demand mathematical tasks
3. Variety of assessment-task formats
4. Appropriate and clear scoring agreements (scoring guide or proficiency rubric)
5. Clarity of directions
6. Academic language

7. Visual presentation

8. Logistics

Common mid-unit assessment and end-of-unit assessment samples designed with the eight common assessment instrument criteria in mind are provided in the next chapter, along with how to establish scoring agreements and calibrate the scoring of student work consistently across members of your team. This is part of **team action 1:** develop high-quality common assessments for the agreed-on essential learning standards.

Visit **go.SolutionTree.com/MathematicsatWork** for free reproducible versions of tools and protocols that appear in this book, as well as additional online only materials.

CHAPTER 3

Sample Common Mathematics Assessments and Calibration Routines

> By keeping our fingers on the pulse of student learning through common formative assessments, we can gain useful information to make adjustments in instruction, provide additional time and support, and give students timely and specific feedback that empowers them to move forward in their learning.
>
> —Kim Bailey and Chris Jakicic

Every common mathematics unit assessment, no matter how well it meets the criteria for high-quality design, can still be improved—crafted into a better tool to more accurately and effectively gather evidence for you and your students as to which essential learning standards students have learned and which they have not learned *yet*.

For some mathematics units or modules, your teacher team may want to create more frequent common assessments to use during the unit. Sometimes, due to the intent of the essential learning standard, common mid-unit assessments (think quizzes, checks for understanding, or exit tickets) may not contain a variety of assessment formats and may use a different balance of higher- to lower-level-cognitive-demand tasks. There are times when your team may also agree to use a performance task or project to assess the learning of students.

This chapter provides various examples of grade-level common assessments and possible team scoring agreements, and explores routines for calibrating your team's scoring of student assessments.

Sample Common Mid-Unit Assessments

Figure 3.1 (page 48), figure 3.2 (page 49), and figure 3.3 (page 50) represent sample common mid-unit assessments for grade 1, grade 5, and grade 8. Select the sample

TEACHER *Reflection*

Describe how you currently use a performance assessment or project based on aligned essential learning standards to strengthen your questions and outcomes.

Currently, how do you and your colleagues agree to score the performance task or project?

How do you identify the purpose of the performance task or project and communicate that purpose to create meaning for the student?

_____
_____
_____
_____
_____
_____
_____

closest to your grade level and determine if the common assessment will provide the information you need to target additional engagement and support for students still not proficient with the standard once they receive your feedback on the common mid-unit assessment.

Directions: Decide if this common mid-unit assessment provides the information your team needs to determine if students have learned the essential learning standard. Will the results give your team the targeted information it needs to best re-engage students in learning when it gives additional time and support?

Name: _____ Date: _____

Grade 1 Place Value Check
I can compare two-digit numbers.

1. In each box, circle the largest number. 3 points _____

21 19	37 60	42 46

2. Circle all the numbers **less than 40**. 2 points _____

   31   52   16   40   68   25

3. Draw a picture using tens and ones to show why **25 is less than 64**. 2 points _____

4. Write >, =, or < in each circle to compare the numbers. 4 points _____

   a. 35 ◯ 38   c. 12 ◯ 21

   b. 80 ◯ 40   d. 91 ◯ 91

5. Carmen says 53 is less than 39 because 3 ones is less than 9 ones. 2 points _____
   Is Carmen correct? Explain your thinking.

   I get it!   I get some of it.   I need help.

Figure 3.1: Grade 1 common mid-unit assessment place value sample.

*Visit **go.SolutionTree.com/MathematicsatWork** for a free reproducible version of this figure.*

Directions: Decide if this common mid-unit assessment provides the information your team needs to determine if students have learned the essential learning standard. Will the results give your team the targeted information it needs to best re-engage students in learning when giving them additional time and support?

Name: _____ Date: _____

Grade 5 Volume Check
I can solve problems related to volume.

For each question, show how you know your answer is correct.

1. A rectangular prism is shown. What is the volume in cubic feet?   1 point _____

   = 1 cubic foot

2. Find the volume of the rectangular prism.   2 points _____

   3 feet, 2 feet, 10 feet

3. Shahira bought an aquarium (a tank for fish) in the shape of a rectangular prism. The tank is 30 inches long and 15 inches tall. It has a width of 20 inches. Shahira wants to fill the aquarium with water. What is the largest volume of water she can put in the tank?   2 points _____

4. Cory made a box in the shape of a rectangular prism. Cory's box has a width of 5 inches, a length of 16 inches, and a volume of 320 cubic inches. What is the height of Cory's box?   2 points _____

5. Amanda is finding the volume of a cube. The cube has a side length of 4 inches. Amanda says the volume is 16 cubic inches. Is she correct? Explain how you know.   2 points _____

- -

Reflection: Circle the sentence that best explains your understanding of solving problems related to volume. Explain your choice.

I can do this with some help.   I can do most of this.   I've got this!

Figure 3.2: Grade 5 common mid-unit geometry assessment sample.

*Visit **go.SolutionTree.com/MathematicsatWork** for a free reproducible version of this figure.*

Directions: Decide if this common mid-unit assessment provides the information your team needs to determine if students have learned the essential learning standard. Will the results give your team the targeted information it needs to best re-engage students in learning when giving the students additional time and support?

Name: _____ Date: _____

Grade 8 Solving Check
I can solve linear equations.

Solve each linear equation. Show how you know each answer is correct.

1. $x + 2x - \frac{1}{2}x = 15$ 　　　　　　　　　　　　　　　　　　　2 points _____

2. $-4(2x - 0.5) = 10x - 1$ 　　　　　　　　　　　　　　　　　　　3 points _____

3. $\frac{2}{5}(x + 10) + 3 = 2 + 2(\frac{1}{5}x + 5)$ 　　　　　　　　　　　　　3 points _____

4. $15x - 3x = -\frac{3}{4}x$ 　　　　　　　　　　　　　　　　　　　　3 points _____

5. Charlie solves the following equation, but he makes a mistake. 　　3 points _____
   Circle the line where the error occurred and complete the solution correctly.

$$14x - 3 - 4x = 14 - \frac{1}{3}(6x - 3)$$
$$10x - 3 = 14 - \frac{1}{3}(6x - 3)$$
$$10x - 3 = 14 - 2x - 1$$
$$10x - 3 = 13 - 2x$$
$$12x = 16$$
$$x = 1\frac{1}{3}$$

- -

Reflection: Circle the sentence that best explains your understanding of solving linear equations. Explain your choice.

　　　I can do this with some help.　　　I can do most of this.　　　I've got this!

Figure 3.3: Grade 8 common mid-unit assessment solving linear equations sample.

*Visit **go.SolutionTree.com/MathematicsatWork** for a free reproducible version of this figure.*

You and your colleagues might also decide to use a common mid-unit assessment exit slip related to a recent student learning target. In preK or kindergarten, your common mid-unit assessments might occur at stations or a kidney bean–shaped table as you observe student thinking and use a recording tool to document students' discussions and answers. Other times, you might create a pencil-and-paper assessment along with a script all teachers can follow when administering the assessment.

Sometimes in the primary grades, teachers may give rolling assessments throughout a unit or a couple of units. Rolling assessments are often given initially during a team-designated window of time and then again, as needed, when students show they may have learned the standard. Alternatively, the team may have several windows of time to give the assessment to monitor student growth and learning. Figure 3.4 (page 52) shows a kindergarten assessment that can be used as a rolling assessment. Since they must assess students one-on-one, teachers may assess a few students a day over the course of a week to gather the information. The team decides whether to assess in one day or use a short window of time. Notice the specificity of the assessment (especially in naming the number of objects to use across the team for each assessment), the rubric, and the scoring sheet, which documents student thinking in real time. (See additional examples in the *Mathematics Unit Planning in a PLC at Work, Grades PreK–2* book; Schuhl, Kanold, Deinhart, et al., 2021.)

In addition to creating the common mid-unit assessments, your teacher team uses the assessment *results* to learn more about student thinking, reasoning, and learning.

Common mid-unit assessments given during a unit should focus on one or two essential learning standards that provide feedback related to student learning along a progression of the essential learning standards. This feedback, in turn, prepares your students for the common end-of-unit assessment.

Assessments given during the unit allow students the opportunity to re-engage in learning essential standards before the end-of-unit assessment. Ideally, you and your colleagues will work to erase inequities in learning as soon as possible as you ensure more students learn the essential learning standards while the mathematics unit movement through the pacing calendar progresses.

TEACHER *Reflection*

You can use these questions from figure 2.1 (p. 18) to help your analysis as you evaluate the quality of each mathematics common end-of-unit assessment.

1. How well does each mathematics task align to the essential learning standards?

2. Is there a balance of lower- and higher-level-cognitive-demand tasks?

3. Is there enough variety between selected-response and constructed-response task formats?

4. How realistic and informative are the scoring agreements? Do you need to clarify the assigned point value or score for each mathematics task on the assessment?

5. Are the directions clearly visible?

6. Are the vocabulary and notations throughout each assessment appropriate to the grade level or course?

7. Is there enough space for students to show their work?

8. How much time is necessary for students taking this assessment? Should you include tools (manipulatives or technology) or not? What other logistics considerations need to be made related to how the assessment is given?

_____
_____
_____
_____
_____
_____
_____
_____

	Kindergarten Essential Learning Standards: I can count to tell how many. I can write numbers 0–20.
Materials	Twenty objects (such as buttons, paper clips, or cubes), white paper and a pencil (for students to write numbers), teacher recording sheet
Logistics	This should be done as an interview. Students are asked to count sets of objects and represent the quantity with a written numeral. Students will demonstrate *one-to-one correspondence* when counting each object and *cardinality* when they know the last number counted represents the number of objects. Note: Ask the student the questions using their home language. You can use a translation app to help you. Students should count in their preferred language. A numeral that is reversed is still correct if it resembles the numeral.
Tasks	For each of the following tasks, rate the student using the rubric that follows. For each question, students can receive a score of 3, 2, or 1. Additional questions are needed to determine if a student should earn a score of 4 (see the rubric). Present the student with a row of 16 buttons. Say: • How many buttons are there? (The student should say the final number.) • Please write that number for me on this piece of paper. Present the student with a 2 × 5 rectangular array of 10 cubes. Say: • How many cubes are there? (The student should say the final number.) • Please write that number for me on this piece of paper. Present the student with a circle of 7 buttons. Say: • How many buttons are there? (The student should say the final number.) • Please write that number for me on this piece of paper. Present the student with a scattered arrangement of 14 paper clips. Say: • How many paper clips are there? (The student should say the final number.) • Please write that number for me on this piece of paper. Extension: Provide the student with more than 23 objects. Ask them to arrange the objects in any way they choose and count them and record the number. Repeat the same activity but with 32 objects.
Rubric	
4 Exceeds	• Student counts accurately more than 20 objects using one-to-one correspondence regardless of the formation. • Student shows a system for keeping track of the count. • Student says the correct number for the final count (cardinality). • Student writes the correct numeral for the final count without reversals.
3 Got it!	• Student counts accurately using one-to-one correspondence within 20 regardless of the formation. • Student shows an efficient way to keep track of the count. • Student says the correct final numeral in the count (cardinality). • Student writes the correct numeral for the final count.
2 Getting there	• Student makes an error in pairing each object with one and only one number name (one-to-one correspondence). • Student has difficulty keeping track of objects counted (counting one object more than once or skipping objects). • Student says the number they count (cardinality). • Student writes the number counted and may switch the digits in the tens and ones places.
1 Not there yet	• Student is unable to pair each object with one and only one number (one-to-one correspondence). • Student cannot keep track of objects while counting. • Student does not recognize the last number name said is the number of objects counted (cardinality). • Student is unable to write the correct number associated with the count.

Teacher Recording Sheet (one-to-one correspondence, cardinality shown with a final count, and written number)																			
Student name	Counting 16 in a row			Counting 10 in an array			Counting 7 in a circle			Counting 14 scattered			Counting 23			Counting 32		
	1-1	Count	Write	1-1	Count	Write	1-1	Count	Write	1-1	Count	Write	1-1	Count	Write	1-1	Count	Write

Figure 3.4: Kindergarten counting (one-to-one correspondence and cardinality) and written numbers common assessment sample.

Visit **go.SolutionTree.com/MathematicsatWork** *for a free reproducible version of this figure.*

TEAM RECOMMENDATION

Create Common Mid-Unit Assessments

- Create shorter common assessments to use during a unit that address one or two essential learning standards.
- Use the assessment instrument quality evaluation rubric (figure 2.1, page 18) to determine the quality of the common mid-unit assessments.

How will you, your colleagues, and your students learn from the results of the common mid-unit assessments? What will you and your colleagues do before the unit ends if students have not yet learned the essential standards assessed?

Common assessments at the end of each unit are also needed to check on any student learning growth from the common mid-unit assessments as well as provide an opportunity for students to demonstrate learning of the standards and make choices regarding their solution pathways. The following are examples of common end-of-unit mathematics assessments that measure student learning in a unit, while also providing information you and your students can use formatively during the next unit.

Sample Common End-of-Unit Assessments

Consider the common end-of-unit assessment examples in figure 3.5 (page 54) and figure 3.6 (page 58). The common end-of-unit assessment in figure 3.5 aligns to the grade 4 fractions unit essential standards in figure 2.3 (page 25).

The end-of-unit assessment in figure 3.6 aligns to the high school essential learning standards for the linear functions unit in the online reproducible "Unit Samples for Essential Learning Standards in Mathematics." (Visit **go.SolutionTree.com/MathematicsatWork** or scan the QR code at the end of the chapter for this free reproducible.)

Choose either the fourth-grade assessment or the high school assessment. As a collaborative team, answer the questions, and then give the exam a high-quality score using the assessment instrument quality evaluation rubric in figure 2.1 (page 18). All 4s in each criterion will result in a total score of 32.

Name: _____ Date: _____

Grade 4 Fractions Unit

I can explain why fractions are equivalent and create equivalent fractions.

1. Circle the two models that show equivalent fractions. Write the equivalent fractions.

    2 points _____

    $\bigcirc = \bigcirc$

2. Write two fractions that are equivalent to $\frac{6}{10}$.

    2 points _____

    $\frac{6}{10} = \bigcirc = \bigcirc$

3. Write two fractions that are equivalent to $\frac{10}{4}$.

    2 points _____

    $\frac{10}{4} = \bigcirc = \bigcirc$

4. Keisha says that $\frac{1}{2}$, $\frac{2}{4}$, and $\frac{3}{6}$ can all be worth the same amount.

    2 points _____

    ‣ Do you agree with her? Circle yes or no: YES   NO

    ‣ Use sketches, numbers, and/or a combination of these to explain your thinking.

I can compare two fractions and explain my thinking.

5. Fill in the circle with <, >, or = to compare the two fractions.

    1 point _____

    $\frac{1}{2} \bigcirc \frac{2}{3}$

6. Fill in the circle with <, >, or = to compare the two fractions.   1 point _____

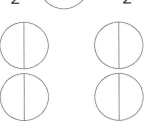

7. Fill in the circle with <, >, or = to compare the two fractions.   6 points _____

$\frac{5}{5}$ ◯ 1 $1\frac{2}{3}$ ◯ $3\frac{1}{2}$ $\frac{4}{5}$ ◯ $\frac{3}{2}$

$\frac{7}{4}$ ◯ 1 $\frac{5}{8}$ ◯ $\frac{3}{4}$ $2\frac{2}{5}$ ◯ $\frac{9}{4}$

8. Mark says $\frac{1}{4}$ of his candy bar is smaller than $\frac{1}{5}$ of the same candy bar.   2 points _____
   He draws the following picture to explain his answer.

   | $\frac{1}{4}$ | | | |

   | $\frac{1}{5}$ | | | | |

Mark's answer is wrong. Explain why Mark's answer is not correct, and then use pictures and words to explain the correct answer.

9. Explain why $\frac{4}{5}$ is greater than $\frac{2}{3}$. Be sure to include a picture in your explanation.   2 points _____

Figure 3.5: Sample grade 4 fractions end-of-unit assessment.   continued →

I can add and subtract fractions and show my thinking.

10. Complete the model and write the matching equation.   3 points _____

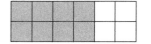 + 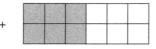 =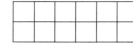

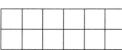

11. Complete the model and write the matching equation.   3 points _____

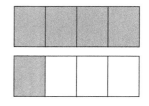 − 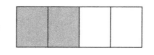 =

 −

12. Show two different ways to add fractions that sum to $\frac{5}{8}$.   2 points _____

$$\frac{5}{8} = \qquad\qquad \frac{5}{8} =$$

13. Find each sum or difference. Show how you know your answer is correct.   4 points _____

$$2\frac{2}{4} + 5\frac{3}{4} = \bigcirc \qquad\qquad 4\frac{1}{3} - 1\frac{2}{3} = \bigcirc$$

14. Charlie uses the following number line to show how to find $1\frac{2}{5} - \frac{3}{5}$. He made a mistake. Find his mistake and then find the correct answer to $1\frac{2}{5} - \frac{3}{5}$ and show how to find it on the number line.

    3 points _____

I can multiply a fraction by a whole number and explain my thinking.

15. Marta says the picture on the number line shows $\frac{1}{6} + \frac{1}{6} + \frac{1}{6} + \frac{1}{6} = \frac{4}{6}$. Jenny says Marta is correct and that the picture on the number line also shows a multiplication equation. Write the multiplication equation Jenny sees in the picture.

    2 points _____

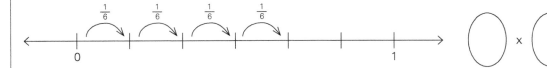

16. Find each product.

    3 points _____

    $$7 \times \frac{1}{8} = \qquad 4 \times \frac{3}{4} = \qquad 5 \times \frac{1}{3} =$$

17. Complete the equation and show how you know your answer is correct.

    2 points _____

    $$2 \times \frac{5}{6} = \bigcirc \times \frac{1}{6}$$

I can solve word problems involving fractions.

18. Mark has five water bottles. Each water bottle holds one liter and is $\frac{2}{3}$ full. He pours them all into an empty water cooler. How many liters of water are in the water cooler now? Show how you know your answer is correct.

    2 points _____

19. Jack is making pancakes. His mom said, "Add flour until the batter is hard to stir." First, he adds $1\frac{3}{4}$ cups of flour, and then he adds $\frac{2}{4}$ cup flour. Finally, the batter is hard to stir. How much flour did Jack add to the pancake batter? Show how you know your answer is correct.

    2 points _____

20. After school, Carli goes for a run. She runs $1\frac{1}{2}$ miles, takes a quick break, and then finishes her run. Altogether, she runs $3\frac{1}{2}$ miles. How far did Carli run after her break? Show how you know your answer is correct.

    2 points _____

Visit go.SolutionTree.com/MathematicsatWork for a free reproducible version of this figure.

Name: _____ Date: _____

Algebra 1 / Integrated Mathematics I Linear Functions Unit

I can graph linear functions and interpret their key features.

1. Graph $f(x) = \frac{2}{3}x - 6$ and identify the slope, y-intercept, and x-intercept. 4 points _____

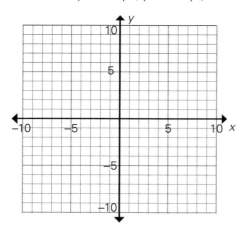

Slope: _____

y-intercept: _____

x-intercept: _____

2. Use the graph of the following function to determine $f(2)$. 1 point _____

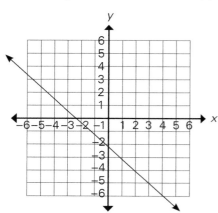

$f(2) = $ _____

3. Graph the line described by the linear equation $x - 6y = 18$, and identify the x- and y-intercepts. 3 points _____

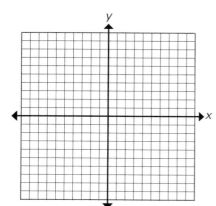

y-intercept: _____

x-intercept: _____

4. Determine the rate of change, y-intercept, and x-intercept of the following function.   3 points _____

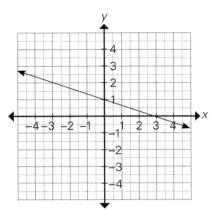

Rate of change: _____

x-intercept: _____

y-intercept: _____

5. Determine the y-intercept of each linear function.   2 points _____

Function A

x	0	5	10	15
y	15	5	–5	–15

y-intercept of function A: _____

Function B

$f(x) = 15x + 13$

y-intercept of function B: _____

I can calculate and explain the average rate of change (slope) of a linear function.

6. Determine the slope of each linear function.   2 points _____

Function A

x	0	2	6	10
y	6	7	9	11

Slope of function A: _____

Function B

$f(x) = -x + 2$

Slope of function B: _____

7. Determine the slope of the line shown in the following graph.   1 point _____

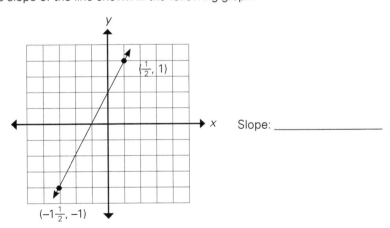

Slope: _____

Figure 3.6: Sample algebra 1 / integrated mathematics I end-of-unit assessment.   continued →

8. The following two linear functions show the amount of money earned for each hour worked at two different stores. Use the slope of each function to explain at which store you would earn more money per hour.

3 points _____

Groceries R Us

Number of hours worked (x)	2	3	5	10
Amount of money earned (y)	18.50	27.75	46.25	92.50

Buy It Here

$9.5x - y = 0$

I can recognize situations with a constant rate of change and interpret the parameters of the function related to the context.

9. The function $f(x) = 40x$ describes how far from home Selena is as she drives from Dallas to Miami. Which graph best represents the function? 2 points _____

A.
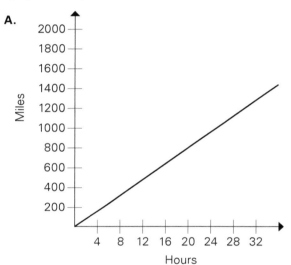

B.
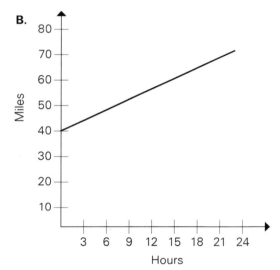

C.
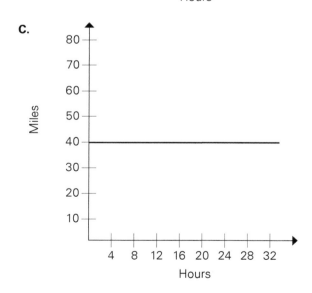

D.
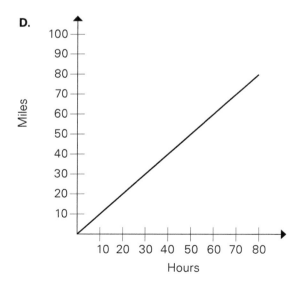

Using the information from your preceding answer, estimate how far from home Selena is in 8 hours.

10. The water level of a river is decreasing at a constant rate every day. The situation is represented by $w(d) = 34 - 0.54d$, where w is the water level in feet after d days.   4 points _____

    a. Determine the rate of change described in $w(d)$, and explain what it means in this context.

    b. Determine the y-intercept for $w(d)$, and describe what it means in this context.

11. When Briana has her picture taken, the photographer charges a $10 sitting fee and $6 for each sheet of pictures she purchases. This can be modeled using the function $f(x) = 6x + 10$, where x represents the number of sheets of pictures purchased. Briana has enough money to buy as many as 10 sheets of pictures.   3 points _____

    a. Determine the constant rate of change in this context.

    b. Determine the domain of this function given the constraints on Briana's amount of money.

    c. Determine the range of this function given the constraints on Briana's amount of money.

12. Martha drives from her house to school. She drives 5 miles, and it takes her 15 minutes.   2 points _____

    a. Determine Martha's average rate of speed.

    b. If you graphed Martha's distance traveled (miles) over time (minutes), will it be a linear function that matches Martha's average rate of speed? Explain why or why not.

I can construct linear functions given a graph, description, or two ordered pairs.

13. Don opens a savings account with $300. Each month, he will add $50 to the account.   2 points _____

    a. Write a function to model this situation where x is the number of months money has been added to the account.

    b. Use your function to determine how long it will take Don to save $1,050.

continued →

14. Brian has 64 flowers for a big party decoration. In addition, he is planning to buy some flower arrangements that have 18 flowers each, **if needed**. All the arrangements cost the same. Brian is not sure yet about the number of flower arrangements he wants to buy, but he has enough money to **buy up to 5 of them**.

    2 points _____

    a. Choose the function that best describes how many flowers Brian has for party decorations. Let x represent the number of flower arrangements Brian buys.

    A. $f(x) = 18x + 64$

    B. $f(x) = 64x + 18$

    C. $f(x) = 64 + x$

    D. $f(x) = 5x + 64$

    b. Choose a reasonable **domain** for the function, keeping in mind that Brian may need to buy additional arrangements.

    A. {0, 1, 2, 3, 4}

    B. {0, 1, 2, 3, 4, 5}

    C. {1, 2, 3, 4, 5}

    D. {5}

15. The following graph shows the total cost of tickets (y) for a given number of rides (x) at a fair.

    4 points _____

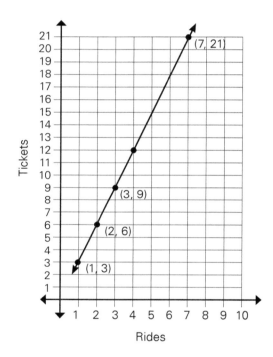

    a. Write a function showing the cost for x rides.

    b. Explain whether or not the model is accurate when drawn with a solid line.

    c. Determine the largest number of rides James can go on if he has $35 to spend on ride tickets. Show how you know your answer is correct.

16. The cost of a prepaid cell phone plan from Cricket includes an activation fee plus a cost per minute. The total costs are represented in the following table.

    5 points _____

    Cricket

Minutes Used (t)	Total Cost $f(t)$
0	$7.50
1	$7.65
2	$7.80
3	$7.95
4	$8.10

    T-Mobile also charges an activation fee plus a rate per minute. The total cost for T-Mobile's plan when t minutes are used is given by $g(t) = 10 + 0.10t$.

    a. Write a function, $f(t)$, to show the total cost of the Cricket cell phone plan based on t minutes of usage.

    b. Which company has the higher activation fee? Provide a mathematical justification for your answer.

    c. Which company charges the most per minute of usage? Provide a mathematical justification for your answer.

17. Josh is planning a large anniversary party for his grandparents. He is deciding how many square tables to put together in a row to seat people. In his following drawings, each table is represented by a square, and each person is represented by a triangle. 5 points _____

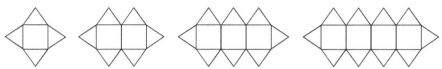

As the drawings show, Josh can seat 4 people around one table and 6 people around two tables put together.
   a. Complete the table to show how many people, p, can sit around t tables put together in a row.

Number of tables (t)	1	2	3	4	5
Number of people (p)					

   b. Write an equation that can determine how many people, p, can sit around a given number of tables, t.

   c. Use your equation to determine how many people can sit around 12 tables.

I can graph piecewise functions, including step and absolute value functions.

18. The graph represents the amount of water in Pam's rain barrel. Is each of the following a possible interpretation of the graph? Check Yes or No for each interpretation given below. 3 points _____

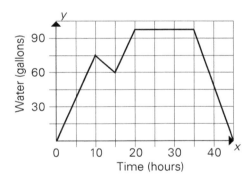

   a. The barrel starts off empty.
      ☐ Yes   ☐ No
   b. Pam uses water from the barrel to water her plants after 35 hours.
      ☐ Yes   ☐ No
   c. It rains for the first 10 hours.
      ☐ Yes   ☐ No

19. Given the absolute value function $g(x) = 3|x| - 5$: 4 points _____

   a. Graph the function on the coordinate grid to the right.

   b. Identify the **vertex** _____.

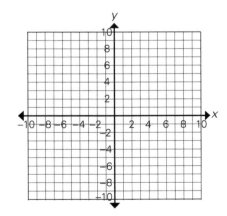

continued →

20. The front of a camping tent can be modeled by the function $f(x) = -1.4|x - 3| + 4$ where the x- and y-axes are measured in feet. Using graphing technology, find the **maximum** height possible for the tent.

2 points _____

21. A bike rental company is having a summer special. The following equations show the prices charged for each hour of rental.

7 points _____

$$f(x) = \begin{cases} 10, & 0 < x < 1 \\ 20, & 1 \le x < 2 \\ 30, & 2 \le x < 3 \\ 40, & 3 \le x < 4 \\ 50, & 4 \le x \le 8 \end{cases}$$

a. Graph the cost of the bike rental for the amount of time the company rents the bike.

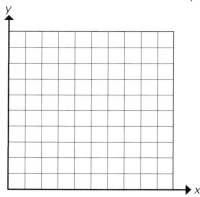

b. Based on the graph, how much will it cost to rent the bike for 3 hours?

c. If you have a budget of $20, determine the maximum amount of time you can rent the bike.

d. What is the maximum length of time possible for the bike rental?

*Visit **go.SolutionTree.com/MathematicsatWork** for a free reproducible version of this figure.*

Did you rate the grade 4 fractions assessment as a perfect 32? Like any assessment, it can be improved. For example, this common assessment is most likely too comprehensive for the time allowed in class.

The grade 4 fractions assessment seems reasonably balanced in terms of higher- and lower-level tasks, but at times may not leave enough space for student work. While the assessment clearly identifies the scoring agreements, there is not quite enough detail on the type of student work needed in explanations to be considered proficient.

If you serve on a fourth-grade team, you may also find some additional or replacement mathematical tasks you feel better meet the essential learning standards listed. That said, this is a strong assessment and certainly one that you can modify to be made even more productive for both teams and students. Strengthening common assessments is part of the professional work of your teacher team every year.

Did you rate the algebra 1 / integrated mathematics I linear functions assessment as a perfect 32? Like the grade 4 fractions example, keep in mind any assessment can be improved. For example, this high school common assessment in figure 3.6 is also too comprehensive for the time allowed in class.

The algebra 1 / integrated mathematics I common assessment seems reasonably balanced in terms of higher- and lower-level tasks, but again, like the fourth-grade assessment in figure 3.5 (page 54), may not leave enough space for student work. You might consider the clarity of directions for a single mathematics task or a group of task questions. Could you improve the directions for clarity?

If you serve on a high school team, you may also find some additional or replacement mathematical tasks you feel better meet the essential learning standards listed. That said, this is a high-quality assessment and one your team can modify. The review process for improving your assessments is part of the professional work of your teacher team every year. (Visit **go.SolutionTree.com/MathematicsatWork** for additional unit assessment samples. See the QR code that appears at the end of the chapter.)

You and your colleagues should take the time to examine all the current common mathematics unit assessment instruments you use with your students to determine strengths and weaknesses.

TEACHER *Reflection*

Compare the unit examples provided to your team's current unit mathematics assessments.

What are the strengths of your mathematics assessments, and in what areas do your assessments need improvement?

Do your colleagues agree with you? Why or why not?

_____
_____
_____
_____
_____
_____
_____
_____
_____
_____

Once you establish your baseline common assessments, you can examine sample online assessments or released items from state, provincial, or national assessments to strengthen and improve your baseline common assessment instruments each school season. Writing and improving high-quality common unit assessments should be an ongoing aspect of your PLC culture. If you are a singleton teacher, consider using released state, provincial, or national mathematics assessment tasks as a way to compare your students against students across a state or province or the nation for discussions with members of your vertical team.

It is equally important to reach team agreement for scoring each mathematics task on your common assessments. The score students receive when taking your common unit assessments should be based on the complexity of reasoning and the necessary work students must show in order to receive either full or partial credit.

Collaborative Scoring Agreements

Recall the Appropriate and Clear Scoring Agreements section in chapter 2 (page 32). Consider these questions: Are the scoring agreements used for every task clearly stated on your assessment? Does the number of total points in a scoring guide or a proficiency rubric level make sense based on the complexity of reasoning for each task or essential learning standard?

Most important, are the points assigned to each task or the levels of proficiency in a proficiency rubric appropriate and agreed on by each team member? Remember, if you and your team use multiple-choice tests, you must also require students to show their work in order to provide accurate and useful scoring feedback using your team-designed rubrics.

In this section, you explore another important step for your team's assessment work. Once you establish your common unit mathematics assessments, you and your colleagues work together and find agreement on scoring those assessments. You consider the demonstrations of learning required for each mathematics task on the assessments and then calibrate your interpretation of student work using your scoring agreements.

Your teacher team's work will depend on whether you use scoring guides with a predetermined point value for proficiency expected (often based on a percentage correct) or proficiency rubrics for each essential learning standard section of the assessment (proficiency is often a score of 3 or higher on a four-point scale) for team and student feedback.

Scoring Guides

In figures 3.7 and 3.8 (page 68), the teacher team placed point values for each mathematics task or question on the assessment. How did the team decide the overall scoring procedures for the common end-of-unit assessment? If you are a singleton teaching an AP course, you may want to consult released exam questions to determine how to allocate points for scoring on your assessments.

Figures 3.7 and 3.8 provide the *scoring guide* the teacher team created for each of the common end-of-unit assessments from figures 3.5 (page 54) and 3.6 (page 58). The scoring guide assigns points to each task with an explanation for how students earn full or partial credit. Choose the assessment closest to your grade level and reflect on the scoring system this team decided to use for each mathematics task and each standard. Determine if you would accept it as is or change it based on your understanding of the essential learning standards.

In the two samples shown in figures 3.7 and 3.8, teacher teams established proficiency for each essential learning standard on the assessment by setting total-points-correct values and expectations for tasks within the essential learning standard.

This type of proficiency setting requires acceptance that if a student meets the total number of points correct (for example, nine of twelve), then the student has demonstrated proficient learning for that standard. Another approach for scoring student assessments is the preparation and use of proficiency rubrics.

Grade 4 Fractions End-of-Unit Assessment Scoring Guide

1. I can explain why fractions are equivalent and create equivalent fractions.

Question	Total Points	Scoring Guidelines (Students must score 6 out of 8 points for proficiency.)
1	2	1 point per correct response (circle first two models: $\frac{3}{4} = \frac{6}{8}$ if they reference the shaded part and $\frac{1}{4} = \frac{2}{8}$ if they reference the unshaded part of the models)
2	2	1 point per correct equivalent fraction to $\frac{6}{10}$
3	2	1 point per equivalent fraction to $\frac{10}{4}$ (includes mixed numbers)
4	2	1 point for yes. However, student can earn one point total if they say says no because the wholes might be different sizes. 1 point if student shows the three fractions are equivalent using the same-sized whole

2. I can compare two fractions and explain my thinking.

Question	Total Points	Scoring Guidelines (Students must score 9 out of 12 points for proficiency.)
5	1	1 point for <
6	1	1 point for =
7	6	1 point for each correct answer = < < > < >
8	2	1 point for noticing Mark drew different-sized wholes 1 point for explaining that $\frac{1}{5} < \frac{1}{4}$
9	2	1 point for pictures showing $\frac{4}{5}$ and $\frac{2}{3}$ 1 point for explaining how the pictures show $\frac{4}{5} > \frac{2}{3}$

3. I can add and subtract fractions and show my thinking.

Question	Total Points	Scoring Guidelines (Students must score 11 out of 15 points for proficiency.)
10	3	1 point for correct shading in final model ($\frac{14}{12}$ shaded) 1 point for correct addends ($\frac{8}{12} + \frac{6}{12}$) 1 point for correct sum ($\frac{14}{12}, \frac{7}{6}, 1\frac{2}{12}$, or $1\frac{1}{6}$)
11	3	1 point for correct shading in the final model ($\frac{3}{4}$ shaded) 1 point for correct fractions on left side of equation ($\frac{5}{4} - \frac{2}{4}$) 1 point for correct difference ($\frac{3}{4}$)
12	2	1 point for correct sum to $\frac{5}{8}$ (such as, $\frac{5}{8} = \frac{1}{8} + \frac{1}{8} + \frac{1}{8} + \frac{1}{8} + \frac{1}{8}$) 1 point for correct sum to $\frac{5}{8}$ different from the first (for instance, $\frac{5}{8} = \frac{2}{8} + \frac{3}{8}$)
13	4	1 point per correct answer ($8\frac{1}{4}, 2\frac{2}{3}$) for a possible total of 2 points 1 point per correct justification for a possible total of 2 points
14	3	1 point for correctly identifying the error (add instead of subtract) 1 point for correct answer ($\frac{4}{5}$) 1 point for showing how to find the answer on the number line

4. I can multiply a fraction by a whole number and explain my thinking.

Question	Total Points	Scoring Guidelines (Students must score 5 out of 7 points for proficiency.)
15	2	1 point for correct factors ($4 \times \frac{1}{6}$) 1 point for correct product ($\frac{4}{6}$)
16	3	1 point for each correct answer ($\frac{7}{8}, 3, \frac{5}{3}$ or $1\frac{2}{3}$)
17	2	1 point for correct answer in equation (10) 1 point for correct explanation

5. I can solve word problems involving fractions.

Question	Total Points	Scoring Guidelines (Students must score 4 out of 6 points for proficiency.)
18	2	1 point for correct answer ($\frac{10}{3}$ liters or $3\frac{1}{3}$ liters) 1 point for justification
19	2	1 point for correct answer ($2\frac{1}{4}$ cups) 1 point for justification
20	2	1 point for correct answer (2 miles) 1 point for justification

Figure 3.7: Sample grade 4 fractions end-of-unit assessment scoring guide.

*Visit **go.SolutionTree.com/MathematicsatWork** for a free reproducible version of this figure.*

Algebra 1 / Integrated Mathematics I Linear Functions End-of-Unit Assessment Scoring Guide

1. I can graph linear functions and interpret their key features.

Question	Total Points	Scoring Guidelines (Students must score 10 out of 13 points for proficiency.)
1	4	1 point for correct graph of $y = \frac{2}{3}x - 6$ 1 point for correct slope: $\frac{2}{3}$ 1 point for correct y-intercept: $(0, -6)$ 1 point for correct x-intercept: $(9, 0)$
2	1	1 point for correct answer: $f(2) = -4$
3	3	1 point for correct graph of $y = \frac{1}{6}x + 3$ 1 point for correct y-intercept: $(0, 3)$ 1 point for correct x-intercept: $(-18, 0)$
4	3	1 point for correct rate of change: $-\frac{1}{3}$ 1 point for x-intercept: $(3, 0)$ 1 point for y-intercept: $(0, 1)$
5	2	1 point for correct y-intercept for function A: $(0, 15)$ 1 point for correct y-intercept for function B: $(0, 13)$

2. I can calculate and explain the average rate of change (slope) of a linear function.

Question	Total Points	Scoring Guidelines (Students must score 5 out of 6 points for proficiency.)
6	2	1 point for correct slope for function A: $\frac{1}{2}$ 1 point for correct slope for function B: -1
7	1	1 point for correct slope: 1
8	3	1 point for slope of Groceries R Us: $9.25 per hour 1 point for slope of Buy It Here: $9.50 per hour 1 point for recognizing you earn more per hour at Buy It Here

3. I can recognize situations with a constant rate of change and interpret the parameters of the function related to the context.

Question	Total Points	Scoring Guidelines (Students must score 8 out of 11 points for proficiency.)
9	2	1 point for correct answer A 1 point for using the graph or $f(x) = 40x$ for answer: About 320 miles
10	4	1 point for rate of change: -0.54 feet/day 1 point for explanation: The water is decreasing $\frac{1}{2}$ foot per day 1 point for y-intercept: $(0, 34)$ 1 point for explanation: Initially, the water level is at 34 feet
11	3	a. 1 point for rate of change: $6 per sheet of pictures b. 1 point for domain: $\{0, 1, 2, \ldots 10\}$ or $0 \leq x \leq 10$ where x is an integer (or whole number) c. 1 point for range: $\{10, 16, 22, \ldots, 70\}$ or $10 \leq y \leq 70$ where y is a multiple of 6 added to 10
12	2	a. 1 point for average rate of change: $\frac{1}{3}$ mile per minute b. 1 point for explaining that it is not a linear function since she did not travel at $\frac{1}{3}$ mile per minute the entire trip (such as, stopped along the way, started slower at first, slowed down at end, and so on)

4. I can construct linear functions given a graph, description, or two ordered pairs.

Question	Total Points	Scoring Guidelines (Students must score 13 out of 18 points for proficiency.)
13	2	a. 1 point for $f(x) = 50x + 300$ (or $y = 50x + 300$) b. 1 point for correct answer: 15 months
14	2	a. 1 point for correct answer A b. 1 point for correct answer B

15	4	a. 1 point for $f(x) = 3x$ (or $y = 3x$)
		b. 1 point for explaining that the domain only includes integers so the graph is not continuous (acceptable if mentioned that the line helps find total cost)
		c. 2 points possible:
		  ▸ 1 point for correct answer of 11 rides (with $2 left over)
		  ▸ 1 point for work or explanation to support the answer

16	5	a. 1 point for the function $f(t) = 7.50 + 0.15t$
		b. 2 points possible:
		  ▸ 1 point for correct answer: T-Mobile has the higher activation fee
		  ▸ 1 point for comparing T-Mobile activation fee of $10 to Cricket of $7.50
		c. 2 points possible:
		  ▸ 1 point for correct answer: Cricket has the higher rate per minute
		  ▸ 1 point for comparing T-Mobile rate of $0.10 per minute to Cricket's $0.15 per minute

17	5	a. 2 points for correctly filling in the table with the values 4, 6, 8, 10, 12 (1 point if at least three are correct) b. 2 points for correct equation: $p = 2 + 2t$ c. 1 point for correct answer: 26 people

5. I can graph piecewise functions, including step and absolute value functions.

Question	Total Points	Scoring Guidelines (Students must score 12 out of 16 points for proficiency.)
18	3	a. 1 point for yes b. 1 point for yes c. 1 point for yes

19	4	a. 3 points for correct graph:
		  ▸ 1 point for vertex at $(0, -5)$
		  ▸ 1 point for left side of graph
		  ▸ 1 point for right side of graph
		b. 1 point for vertex $(0, -5)$

20	2	2 points for sketch and correct answer: 4 feet

21	7	a. 4 points for correct graph, specifically:
		  ▸ 1 point for labels
		  ▸ 1 point for closed points on the left of each "piece"
		  ▸ 1 point for each open point on the right of each piece
		  ▸ 1 point for connecting each piece in horizontal segments
		b. 1 point for correct answer $40
		c. 1 point for correct answer: up to 2 hours (1 hour 59 minutes)
		d. 1 point for correct answer: 8 hours

Figure 3.8: Sample algebra 1 / integrated mathematics I end-of-unit assessment scoring guide.

*Visit **go.SolutionTree.com/MathematicsatWork** for a free reproducible version of this figure.*

Proficiency Rubrics

Some collaborative teams use a proficiency rubric to score student work for essential learning standards. Teachers use the rubric to determine if the student has one of the following for each essential learning standard.

- Minimal understanding
- Partial understanding
- Proficient understanding
- Above-grade-level understanding

The names given to the proficiency levels can vary. At Adlai E. Stevenson High School (2022), the levels are referenced as *developing foundations* (level 1), *approaching proficiency* (level 2), *meets proficiency* (level 3), and *exceeds proficiency* (level 4).

A proficiency rubric can be a good way to measure student progress on a standard over time and during several units of study, as well as within a single unit of study. Proficiency rubrics can be used in preK–12 if you identify the essential learning standards to monitor for each rubric. Instructionally, your teacher team will clarify success criteria for each proficiency level so students more clearly understand what they need to learn.

Figure 3.9 shows an example of a proficiency rubric for the first-grade common mid-unit assessment in figure 3.1 (page 48). Notice that it is not a rubric for each question on the assessment, but rather an overall proficiency rating related to the essential learning standard. Also notice that students cannot earn a level 4 on this assessment since all task questions are at grade level.

Directions: Compare the proficiency scale for the place value check with the assessment in figure 3.1 (page 48). How would you and your teammates interpret each score? What would a student have to show to earn a 1, 2, or 3?

Level 1 Minimal Understanding	Level 2 Partial Understanding	Level 3 Proficient Understanding	Level 4 Above Grade-Level Understanding
Student can sometimes identify the largest number when comparing two numbers.	Student can identify the greater than or less than number and use the correct symbol for comparison, but cannot explain the comparison using place value. Or Student can identify the greater than or less than number and can often use the correct symbol for comparison with a minimal ability to explain the comparison using only ones (not tens as a group).	Student can identify the greater-than or less-than number and use the correct symbol for comparison (with a possible simple mistake for an error) and can explain the comparison using place value language.	Not applicable for this assessment; needs additional data. Ask student to compare three-digit numbers and explain why one number is larger than another using place value language or have student identify a number that is greater than a given number and less than a second number, and write the entire inequality using two inequality symbols.

Figure 3.9: Proficiency rubric for grade 1 common mid-unit assessment sample in figure 3.1.

If you and your colleagues use proficiency rubrics like the one in figure 3.9, it is important to identify the quality of student work expected at each proficiency level as it relates to the standard, rather than inserting points into the scale (for example, avoid saying a student must get ten points to earn a 3 on the rubric). It is also important to determine whether the common end-of-unit assessment is the best place to assess students at level 4 or if students can demonstrate learning at level 4 through formative assessment experiences, through activities during the unit, or with work on a separate level 4 common assessment given after or at a different time than the original common assessment.

On a common end-of-unit assessment, there would be one proficiency rubric per essential learning standard. Figures 3.10 and 3.11 (page 72) show possible proficiency rubrics for the end-of-unit assessments in figures 3.5 (page 54) and 3.6 (page 58).

Grade 4 Fractions End-of-Unit Proficiency Rubrics

1. I can explain why fractions are equivalent and create equivalent fractions.

Level 1 Minimal Understanding	Level 2 Partial Understanding	Level 3 Proficient Understanding	Level 4 Above-Grade-Level Understanding
Student identifies an equivalent fraction with limited denominators such as halves and fourths or fourths and eighths but struggles to model equivalent fractions with less friendly denominators.	Student creates equivalent fractions from a given fraction or model but cannot explain why the fractions are equivalent.	Student creates two equivalent fractions from a given fraction or model and explains why the fractions are equivalent.	Student's explanation of equivalent fractions involves a generalization that can be applied to any fraction, such as "doubling the numerator and denominator always results in an equivalent fraction."

2. I can compare two fractions and explain my thinking.

Level 1 Minimal Understanding	Level 2 Partial Understanding	Level 3 Proficient Understanding	Level 4 Above-Grade-Level Understanding
Student compares two fractions using <, >, or = when given a model and needs support when comparing fractions without a model.	Student compares two fractions using <, >, or = when given a model and has minor errors when comparing without a model.	Student compares two fractions using <, >, or = when given a model and without a model and shows an understanding that fractions can only be compared when they reference the same whole.	Student compares two fractions using <, >, or = with fluent procedures and explains with words that reference benchmarks and reasoning as well as models the comparison of two fractions.

3. I can add and subtract fractions and show my thinking.

Level 1 Minimal Understanding	Level 2 Partial Understanding	Level 3 Proficient Understanding	Level 4 Above-Grade-Level Understanding
Student adds and subtracts fractions with like denominators when given a model, with limited accuracy.	Student adds and subtracts fractions with like denominators when given a model and without a model (with a possible simple mistake for an error). Student may struggle with regrouping mixed numbers when adding and subtracting or can only decompose a fraction one way.	Student adds and subtracts fractions with like denominators (including regrouping mixed numbers if needed), decomposes a fraction more than one way, and shows thinking using models and equations.	Student shows efficient and flexible procedures to add and subtract fractions. Additional data may be needed—student demonstrates procedural fluency routinely without models.

4. I can multiply a fraction by a whole number and explain my thinking.

Level 1 Minimal Understanding	Level 2 Partial Understanding	Level 3 Proficient Understanding	Level 4 Above-Grade-Level Understanding
Student multiplies a fraction $\frac{1}{b}$ by a whole number with a focus on "rules" and makes several errors (for example, multiplies the numerator and denominator by the whole number).	Student multiplies a fraction $\frac{a}{b}$ by a whole number with minor mistakes and struggles to show an understanding of repeated addition with unit fractions and the connection to multiplication of a whole number and fraction.	Student multiplies a fraction by a whole number using models and equations to show thinking. Student demonstrates an understanding of $\frac{a}{b}$ as a multiple of $\frac{1}{b}$.	Not applicable for this assessment; needs additional data. Ask student to multiply a whole number by a mixed number. (Student may rewrite the mixed number as an improper fraction or use the distributive property to multiply the whole number by the mixed number as written.)

Figure 3.10: Example proficiency rubrics for the grade 4 end-of-unit assessment in figure 3.5.   continued →

5. I can solve word problems involving fractions.

Level 1 Minimal Understanding	Level 2 Partial Understanding	Level 3 Proficient Understanding	Level 4 Above-Grade-Level Understanding
Student attempts to draw a picture to represent the word problem, but often misinterprets which operation is needed to solve the problem, and may need support to understand what the problem is asking.	Student solves some of the addition, subtraction, and multiplication word problems involving fractions and shows their thinking with limited accuracy. Student may struggle when the context is more complex and the unknown is at the start or middle.	Student solves a variety of addition, subtraction, and multiplication word problems with varied unknowns involving fractions and shows their thinking using models or equations.	Student shows procedural fluency with solving word problems involving fractions. Additional data may be needed; ask student to create a word problem given the answer or the equation needed to solve the word problem, or ask student to solve problems with multiple steps that require more than one operation.

Algebra 1 / Integrated Mathematics I Linear Functions End-of-Unit Proficiency Rubrics

1. I can graph linear functions and interpret their key features.

Level 1 Minimal Understanding	Level 2 Partial Understanding	Level 3 Proficient Understanding	Level 4 Above-Grade-Level Understanding
Student's early understanding of key features of linear functions results in their inability to correctly graph functions with consistency, and/or given a function or a graph, student cannot identify key features with consistency.	Student possesses a limited understanding of the relationship among the various representations of functions. Student may demonstrate an ability to graph *or* identify some or all key features.	Student demonstrates the ability to graph, interpret, and identify key features of a given function with consistency. Student may make minor errors when completing a task, but those errors do not interfere with their understanding.	Student possesses a relational understanding of the function, key features, and graph. Student works flexibly among representations, demonstrating the ability to correctly identify key features with consistency.

2. I can calculate and explain the average rate of change (slope) of a linear function.

Level 1 Minimal Understanding	Level 2 Partial Understanding	Level 3 Proficient Understanding	Level 4 Above-Grade-Level Understanding
Student possesses a limited understanding of rate of change (slope) that interferes with the ability to calculate slope given a function, table, graph, or real-world scenario. Student may possess the procedural understanding to calculate slope given one or two of these presentations, but they do so with inconsistent results.	Student utilizes procedural understanding to calculate slope given a function, table, or graph but the work frequently contains calculation errors that result in incorrect responses. Or student can consistently calculate slope (procedurally) given a function, table, or graph but the limited relational understanding interferes with the ability to explain how the slope is represented in the other representations, especially when given a real-world scenario.	Student demonstrates a relational understanding of slope given a function, table, graph, or real-world scenario. Student can calculate slope correctly, and any minor errors do not interfere with understanding.	Student possesses a deep conceptual understanding of slope and can calculate slope flexibly given a function, table, graph, or real-world scenario. Student understands the connectedness among various representations and can apply that understanding to reason and make sense of authentic problems.

3. I can recognize situations with a constant rate of change and interpret the parameters of the function related to the context.

Level 1 Minimal Understanding	Level 2 Partial Understanding	Level 3 Proficient Understanding	Level 4 Above-Grade-Level Understanding
Student may recognize situations with rates of change, but struggles to calculate and/or apply knowledge of the rate of change given a real-world scenario. Student responses are incompatible with reasonable solutions for the context of the problem and, thus, expose flaws in understanding.	Student can sometimes calculate the rate of change given a real-world scenario but cannot explain the value in the context of the problem. Or Student can calculate the rate of change and explain that value in the context of the problem but cannot apply that knowledge to solve related problems that extend or expand the scenario.	Student can calculate the rate of change and explain that value in a real-world context with minor errors not interfering with understanding. Or Student can consistently calculate the correct rate of change and explain that value in the context of the problem but is less consistent when applying that knowledge to extend or expand the scenario.	Student calculates the rate of change, explains that value in the context of the problem, and is able to extend or apply that knowledge with a great deal of consistency, flexibility, and accuracy.

4. I can construct linear functions given a graph, description, or two ordered pairs.

Level 1 Minimal Understanding	Level 2 Partial Understanding	Level 3 Proficient Understanding	Level 4 Above-Grade-Level Understanding
Student is unable to write or identify a function that best describes a real-world scenario. They can sometimes produce a correct response using a strategy that is not aligned to the learning target.	Student can identify a function, given choices, that best represents a real-world scenario but is unable to write a function independently. And/or student does not possess the relational understanding necessary to work flexibly among various representations (real-world scenario, table, or graph) of functions modeling real-world scenarios and, therefore, cannot consistently produce correct solutions.	Student correctly writes and identifies a linear function given a real-world scenario, table, or graph. Student possesses a relational understanding that produces correct responses consistently with minor flaws not interfering with understanding.	Student can write and identify a linear function given a real-world scenario, table, or graph. Student possesses a deep conceptual understanding of the content that produces correct responses with a great deal of consistency, flexibility, and accuracy.

5. I can graph piecewise functions, including step and absolute value functions.

Level 1 Minimal Understanding	Level 2 Partial Understanding	Level 3 Proficient Understanding	Level 4 Above-Grade-Level Understanding
Student can sometimes produce the correct response when graphing or interpreting *either* a step function or an absolute value function (but not both). Student's response demonstrates a lack of understanding of how these functions, used to model specific real-world scenarios, are different from the linear functions from this unit.	Student can consistently produce the correct response when graphing or interpreting *either* a step function or an absolute value function (but not both). Or Student can produce the graph of both functions but is unable to use that graph to interpret solutions in a real-world context.	Student can consistently produce correct responses when graphing or interpreting both a step function and an absolute value function. Student is able to use the graph to interpret solutions in a real-world context, but sometimes, minor errors in the graph or interpretations lead to incorrect responses. These minor errors do not interfere with understanding.	Student can consistently produce correct responses when graphing or interpreting a step function or an absolute value function *and* they possess a deep conceptual understanding of how these functions are used to model different real-world scenarios than those used for linear functions. Student response is completely accurate.

Figure 3.11: Example proficiency rubrics for the algebra 1 / integrated mathematics I end-of-unit assessment in figure 3.6.

In the following personal story, author Sarah Schuhl describes her experiences working with teams to make sense of developing proficiency rubrics. Sarah's experience with this high school team is similar to PLC at Work teacher team discussions for elementary and middle school teachers. The proficiency rubric should describe the overall student work necessary to demonstrate each level of proficiency rather than assigning a number of points needed within the rubric for proficiency.

In a PLC at Work, your scoring guides and proficiency rubrics ensure a consistency in the accuracy and depth of teacher feedback to students. New teachers to a grade level or course on your team will appreciate knowing these guidelines.

> **TEACHER** *Reflection*
>
> How can a proficiency rubric show student growth with an essential standard over time?
>
> Does your teacher team currently score your common unit assessments using a scoring guide or a proficiency rubric? Which do you prefer? Why?
>
> _____
> _____
> _____
> _____

Personal Story SARAH SCHUHL

In my work with teams, I have learned that proficiency rubrics are not always easy to create, but they provide rich opportunities for teachers to talk about what students have to know and be able to do to be proficient with the mathematics standards.

At a high school collaborative team meeting with a team transitioning to scoring assessments using proficiency rubrics, the teachers wanted my opinion on their rubric. The rubric did not reference a total number of points needed for the essential learning standard portion of the assessment. Interestingly, the rubric also did not reference student learning levels related to the standard. Instead, the proficiency rubric identified which questions a student had to answer correctly to earn a score of 1, 2, 3, or 4. For example, students earned a 1 if they answered questions 1 and 2 correctly, a 2 if they also answered question 3 correctly, and a 3 if they also answered question 4 correctly. Question 5 was beyond the intent of the standard, so if students answered all five questions correctly, they earned a 4 on the proficiency rubric. Using such a proficiency rubric, what happens if a student only answers question 3 correctly? Or question 5? What happens if a student misses the first question and the rest are perfect?

As we wrestled with the options, it became clear the team was going to need to create a rubric showing the type of learning evidence a student produces for each level, 1–4. We looked at a rubric from ThemeSpark (www.themespark.net) and revised it as needed, noting that a proficient student might still make some simple mistakes and need additional practice, but is not in need of reteaching.

The team also decided to assess level 4 not on the common end-of-unit assessment but rather as a separate test. This would allow them to ask more questions at grade level and make sure students did not spend too much time on questions beyond the intent of the standards at the expense of having time to demonstrate learning on other essential standards. Instead, the team created five questions, some at a low level and others at a high level (but not beyond the intent of the essential mathematics standards). Team members did this only after they made their proficiency rubric, so they could discuss the type of work expected for a student to earn a score of 1, 2, or 3 using the preponderance of evidence produced by the student on that portion of the exam.

Whether you use total points from a scoring guide or proficiency rubric levels, you and your mathematics team should define scoring agreements and calibrate the scoring agreements for all common mid-unit and end-of-unit assessments.

To support your collaborative scoring work, choose one of your current common mid-unit or end-of-unit assessments, preferably one you will be using soon. Use the tool in figure 3.12 to determine how you will score the mathematical tasks on the assessment and how you will know whether students are proficient for each essential standard.

There is no right or wrong way to determine your team scoring agreements. What is important is that you and your colleagues use the same scoring guide for tasks within an essential learning standard or the same proficiency rubric score for each group of tasks aligned to the essential learning standard (consider the complexity of reasoning required by the assessment task or proficiency rubric based on lower- or higher-level cognitive demand).

Reaching team scoring agreements will create a greater likelihood of erasing any student inequities in your individual scoring and grading practices once students take your common assessments. There is also an increased fidelity to your answer to PLC critical question 2 (DuFour et al., 2016): How will we know if students learn it?

That answer lies partially in the hard work you do as a collaborative team to reach scoring consensus with your colleagues.

Assessment Instrument Alignment and Scoring Agreements

Directions: Within your collaborative team, answer each of the following five questions in relation to your team common assessment.

1. Which essential learning standard does each task address, and how do you know that the task aligns to the essential learning standard?

2. What work do you expect students to demonstrate to successfully respond to each task on the assessment and either receive full credit when using a scoring guide or earn a 3 on a proficiency rubric for each task? How might students earn partial credit or demonstrate a partial understanding?

3. Which mathematical practices or process standards does the task develop?

4. If you score with points on a scoring guide, how many total points should you assign to the end-of-unit assessment? If you score using a proficiency rubric, how will you score students as proficient for each essential learning standard?

5. Are there any questions on the test you want to ask differently? If so, how will that affect the point value on a scoring guide or the proficiency rubric score you will assign?

Figure 3.12: Teacher team discussion tool—Assessment instrument alignment and scoring agreements.

Visit **go.SolutionTree.com/MathematicsatWork** for a free reproducible version of this figure.

>
>
> **TEAM RECOMMENDATION**
>
> ## Create Common End-of-Unit Assessments
>
> - Create common assessments to use at the end of each unit that address the two to five unit essential learning standards. Ideally, create these common assessments before the unit is taught.
>
> - Using the eight criteria in figure 2.1 (page 18), the assessment instrument quality evaluation rubric, determine the quality of your common end-of-unit assessments.

As your collaborative team begins the assessment work within a PLC culture, you will pursue building deliberate agreement on the appropriate scoring of student work for each mathematics problem or task on the assessment. The criteria for scoring, as described in this chapter, will help you, your students, and your students' families gain confidence that the score is accurate and provide the necessary feedback for the formative purposes of assessment described in chapter 1 (page 5). Scoring agreements are the first step in your teacher team calibration routines. In addition to agreement on *how* to score the assessments is the need to *calibrate* how assessments are scored with fidelity across members of your team once the common assessments have been administered.

Quality of Mathematics Assessment Feedback

Teacher and student action based on the results of your common assessments hinges on the nature of your consistent feedback to students from teacher to teacher on your team. Consistent feedback will, over time, erase any potential inequities for student learning opportunities.

The ability and willingness of your teacher team to calibrate scoring on common assessments allows for a strong and shared intervention program with student re-engagement in learning and facilitates student goal setting and self-efficacy development.

First, your team scores and calibrates feedback for common unit assessments, and then your students act on the subsequent feedback.

Calibration Routines

Suppose your teacher team gives a common assessment on Tuesday and then plans to review the results with students on Thursday. Unfortunately, life has been hectic and grading your assessments in time seems like an enormous challenge.

If a colleague knows of your situation and offers to grade your common unit assessments for you, how do you respond? Do you feel relieved and grateful for the help and support, or do you panic at the thought that the colleague might not score the assessments the same as you?

In chapter 2 (page 17), you determined common scoring agreements when designing the tasks for each unit assessment. Once you give the common assessment to students, there is an urgent moment when your team meets to calibrate your scoring and receive feedback from each other on your scoring accuracy for the common unit assessments. This is urgent because you need to return the feedback in a timely manner (recall the *T* in FAST feedback means *timely*), and you will need to reduce the potential for wide variability in how you and your team members score similar assessment tasks and questions.

It is in this moment (after students take an assessment) that you can agree on the scoring accuracy of the feedback you give to students. If you do not agree, grading and scoring inequities will occur within the grade level or course. Scoring student work together provides insight into other team members' mathematical thinking and allows you to consider multiple ways you and your students represent their thinking.

Fidelity to the formative feedback process occurs when your collaborative team meets to score student work fairly and consistently as a routine part of practice. To strengthen this practice of consistently scoring student work, your team should practice common scoring of the same tasks at the end of each unit as part of a team calibration routine. As the story from author Sarah Schuhl indicates, the results of your initial calibration work together may be surprising.

> ## Personal Story SARAH SCHUHL
>
> When I first asked our algebra team to collectively score a common unit assessment on linear equations, team members indicated it would be a waste of time. They had written the test together as well as shared how many points each item was worth on the assessment. However, they agreed to try.
>
> We met together and scored a student assessment, only to learn the student would have earned an overall score of D– to B+, depending on which of the five teachers on the collaborative team graded the test. We were shocked!
>
> If we had handed back assessments for reflection and goal setting without calibrating our scoring, some students would have been told they needed intervention and others told they met the essential learning standards, depending on which teacher might have scored the assessment.
>
> This disparity meant that using the same assessment to the same standards is not enough to ensure equitable outcomes for students across each teacher on our team with respect to consistently grading student work on our common assessments and providing feedback from assessments or during instruction.

To view common assessments through more of an equity lens, students should receive similar scores from each team member (within one or two points for the entire end-of-unit exam or the same proficiency rubric score for each essential learning standard). This also translates into similar scores for each essential learning standard on the assessment.

Some more common methods for ensuring scoring agreement routines include the following.

- **Double scoring:** This means one teacher scores exams and then a second teacher also scores the exams, placing both grades on the assessment. When there are variances, a third scorer or grader (usually a teacher on the team or a teacher who knows the mathematics content standards for the unit assessment) may be brought in to also score the exam, or the teachers work to clarify and agree about how to commonly score the essential learning standard or task in question.
- **Calibration:** This is when the teachers agree on the points or rubric to use to score papers and use anchor papers to gain clarity about student work and its interpretation related to proficiency.
- **Inter-rater reliability:** This occurs when two or more teachers score the same paper and examine their results during blind double scoring. Over time, and as teachers engage in discussions about how they each arrived at the final assessment score, they resolve any significant differences and begin to assign student scores with feedback that is reasonably close. Thus, there becomes a greater score reliability across teachers scoring (or rating) the common assessment.

> ## TEACHER *Reflection*
>
> Do you currently double score your unit assessments with a colleague? In other words, do you grade each other's student work? If so, where and when do you get it done?
>
> _____
> _____
> _____
> _____
> _____
> _____
> _____
> _____

When you calibrate scoring as a teacher team, you also calibrate your own understanding of the standards and proficiency expectations, which translates into accurate interpretation of results for teachers and students alike. Additionally, interventions and extensions are more accurately designed when student work has been consistently scored across your team.

If you give a common assessment in a one-on-one interview setting and record student responses (such as in preK or kindergarten), your team may need to film a few of your assessment interviews for the purpose of calibrating your scoring. You can use the videos in a team meeting to calibrate your interpretations of student learning as evidenced in real time. How closely and accurately do you and your colleagues score student evidence of learning?

You can use the teacher team activity in figure 3.13 to facilitate a collaborative scoring conversation around either a recent common end-of-unit assessment your team gave or an upcoming assessment for the end of a unit. In a few instances, you may need to bring in videos from common assessments rather than student papers.

Directions: Choose one other team member and plan to meet for thirty minutes at the end of the day. Each brings six to eight student papers or assessments to grade and score together. Use the following five steps to calibrate your scoring agreements and student feedback.

1. Bring student assessments that you believe will show strong evidence of student learning (proficient or advanced), partial evidence of student learning, and minimal evidence of student learning (in other words, those assessments that might rank high, medium, and low).

2. Blind double score the assessments using your agreed-on scoring agreements, not letting your scoring partner know the score you give the student work or the feedback you would provide.

3. Compare scores and determine if there are any score variances for any standards or areas of the assessments.

4. Examine the mathematical tasks on the assessment that caused the greatest variance, and resolve those issues.

5. Once you calibrate your scoring expectations for student work with a teacher partner, score the rest of the assessments on your own using the feedback you receive from your colleague and the agreements you generate together.

Figure 3.13: Teacher team discussion tool—Equitable scoring and calibration routine.

*Visit **go.SolutionTree.com/MathematicsatWork** for a free reproducible version of this figure.*

Work with your colleagues to determine the best methods for providing students with feedback that promotes the team's desire for continuous student learning. Feedback to students starts with calibrated scoring agreements on both shorter common mid-unit assessments and common end-of-unit assessments. Students know their score will not change if another teacher on the team grades the assessment.

The information the team gleans from each assessment is more useful when it is gleaned by essential learning standard so all stakeholders can determine what students have learned and not learned yet, providing hope for improvement in students' and teachers' minds alike.

Once team members work through the calibration activity in figure 3.13, it is important to continue addressing assessment-scoring calibration together with future common unit assessments. What are your team routines for calibration? You might also consider rotating your scoring partner as you seek a more equitable and calibrated student feedback response on your common assessments.

If you are a singleton, you may need to use the calibration activity in figure 3.13 with other colleagues in your department to ensure calibrated grading vertically, so students know feedback is consistent from one year to the next in mathematics. You might also calibrate by grading released student work from a state (or provincial) or national assessment and comparing your scoring to that of the released examples. Finally, you might engage with a virtual team, in which case you can engage in calibrating your scoring and feedback electronically. It is critical that students experience equity in their scoring feedback for optimal continued learning.

Quality Student Feedback

As your teacher team discusses how to calibrate grading, consider too if a single score, whether based on points from a scoring guide or a level from a proficiency rubric, accurately reflects student learning using the assessment. Beyond the scoring of the assessment, how does your collaborative team ensure the assessment feedback is FAST—fair, accurate, specific, and timely? How is your team working to ensure students can learn from the feedback provided so learning does not stop because of a single score placed on an assessment?

John Hattie, Douglas Fisher, and Nancy Frey (2017) remind us that "what we say to students, as well as how we say it, contributes to their identity and sense of agency, as well as to their success" (p. 203). Further, in Hattie, Fisher, and Frey's (2017) research, there is a significant impact on student learning (an effect size of 0.75) through frequent and high-quality formative feedback on assessments. The end-of-unit assessment is part of that process as students continue to learn throughout the next unit of mathematics.

As an example, imagine a student only provides one solution pathway to solve a problem, and you provide feedback hints on the assessment task that could guide the student to consider another approach, such as, "What related facts do you know that could help you?" or "How could you uswe a graphing strategy?" If you provide feedback to the student well after they have taken the assessment, your feedback will have minimal impact. Students too often forget what they were thinking when they took the exam and tend to start over instead of working to revise their thinking and reasoning—an action that has a greater impact on their learning.

> **TEACHER** *Reflection*
>
> Remember, the use of accurate and specific feedback only helps students if it is also timely. In general, how quickly does a student receive feedback from you on a mid-unit or end-of-unit assessment performance? What are the greatest barriers you face to providing timely feedback, and how can your team improve in this important part of the common mathematics assessment process?
>
> _____
> _____
> _____
> _____
> _____
> _____
> _____
> _____

> **TEAM RECOMMENDATION**
>
> ### Use Calibration Routines for Common Unit Assessments
>
> - Double score common assessments to calibrate feedback to students.
> - Discuss how to give FAST feedback to students within a quick time frame.
> - Determine how you and your colleagues will calibrate scoring and explore effective ways to give students meaningful feedback that is consistent from teacher to teacher.

Thus far, you and your colleagues have determined essential learning standards, created high-quality common assessments with common scoring agreements, and calibrated your scoring of student work on your common mathematics assessments (surfboards). From figure 1.2 (page 12), you have established the conditions for your team and students to learn from the common assessment results (surfing). How will each common mid- and end-of-unit mathematics assessment become a valuable part of the formative learning process for you and your students?

In the next chapter (chapter 4), you will explore how your teacher team might use common assessment results to analyze effective instructional practices and determine students in need of continued learning.

In chapter 5 (page 97), you will examine how to use common assessments with students for continued learning of the essential learning standards.

Then, in chapter 6 (page 121), you will learn about effective Tier 2 interventions for continued student learning.

Each common mathematics assessment is only as valuable as the timely feedback it provides to you and your students and the opportunities it uncovers for targeted interventions and extensions. All of these actions are part of **team action 2:** analyze and use common assessments for formative student learning and intervention.

Visit **go.SolutionTree.com/MathematicsatWork** for free reproducible versions of tools and protocols that appear in this book, as well as additional online only materials.

CHAPTER 4

Teacher Actions in the Formative Assessment Process

The most powerful lever for changing professional practice is
concrete evidence of irrefutably better results.

—*Richard DuFour*

Once the common assessments have been given, how will students and teachers learn from the results? You and your teacher team members learn from analyzing student work together. Which instructional practices proved most effective for student learning? Which students demonstrated learning of the standards and which did not, yet? What will be your team response to the evidence of student learning? This chapter addresses teacher data-analysis and action routines as the fourth criterion for teams using common assessments as part of a formative assessment process (figure 1.2, page 18).

Author and former educator Roland S. Barth (2001) says it best:

> Ultimately there are two kinds of schools: learning-enriched schools and learning-impoverished schools. I've yet to see a school where the learning curves . . . of the adults were steep upward and those of the students were not. Teachers and students go hand in hand as learners—or they don't go at all. (p. 23)

Creating opportunities for your teacher team to collectively examine student learning opens the door for you and your colleagues to take risks and improve your teaching practices—and become a learning-enriched school. You are engaging in the formative assessment process (surfing) by learning about your instructional practices and student thinking. Your teacher team then uses the evidence (or lack of evidence) of student learning to ensure students receive formative feedback and re-engagement opportunities. The formative process for student learning occurs when students make their thinking on each task visible, whether that assessment is taken online or on paper.

The following are a few questions you and your colleagues might use when analyzing and learning from your common assessment results.

- How did each student perform on the common assessment for each essential learning standard?

- How did the results vary across the members of our team? Which instructional strategies worked best and why? What can we learn from each other about improving our instructional strategies and routines?

- As a team, for which essential learning standards on the assessment did the students meet proficiency? What trends in student reasoning emerged from student work?

- Which students need additional time and support to learn specific essential standards? What strategies might we explore to increase student learning for each standard?

Analyzing data together can be unnerving for some team members at first. Jon Yost, former associate superintendent of Sanger Unified School District in California, reminds teachers "data [are] for learning, not for judging" (J. Yost, personal communication, October 5, 2022). The focus when analyzing data is to uncover how well your students are learning, and then identify targeted team interventions and extensions as needed.

You and your colleagues will learn about evidence of student reasoning as well as evidence of your most effective instructional practices as revealed by the student work. You will also learn which students have learned essential learning standards and which have not *yet* learned essential learning standards, and thus are in need of targeted and specific plans for intervention and extension. The data-analysis process will lead to greater transparency and learning across your teacher team.

Jody Guarino, Shelbi Cole, and Michelle Sperling (2022) challenge thinking on the meaning of data in their article "Our Children Are Not Numbers." The authors talk about humanizing data and ask, "How do our current uses of data support or diminish students' agency, identity, and belonging?" and "Is our definition of data broader than numbers and percentages, perhaps including information students give us based on what they know, evidence of their current thinking how they are making sense of things, what they understand, or what they are working toward understanding?" (p. 406). In *Catalyzing Change in Middle School Mathematics*, NCTM (2020) states:

> Conscious and unconscious beliefs exist about what each and every student can learn and do in the mathematics classroom. These beliefs translate into equitable or inequitable practices by either positioning each and every student to be or not to be learners and doers of mathematics. (p. 22)

Thus, your collaborative team considers how to use student assessment data with student work to form an asset-based understanding of what students have learned. You use the knowledge gained from the data to continue growing student learning and simultaneously contribute positively to your students' self-efficacy and mathematics identity.

In a PLC at Work, common assessments are part of a student's learning story rather than their judging, measuring, and recording story. Common assessments are simply tools (surfboards) in the formative learning process (surfing). It is often difficult to know where to start as a teacher team when analyzing student evidence of learning. The three protocols in this chapter are considered *data dialogue* protocols because they illustrate the learning evidence to gather and pose questions for you and your colleagues to discuss. Each protocol includes instructions and a template to record your teacher team discussions and plans. Your teacher team may want to complete your data dialogues electronically so each teacher can input data when the assessments are scored and the entire team can see the data and contribute to recording your team's analysis and next steps.

Analyzing data is more than just looking at numbers. The data dialogue protocols provided highlight a structured routine for your team to more closely examine student learning and make a targeted plan for continued instruction, intervention, and extension. In each protocol, student learning may be revealed from written work or observations from one-on-one assessments scored with student checklists. If your teacher team is using an assessment program to gather student data, those data are often seen only by the assigned mathematics teacher. The dialogue protocols in this chapter provide a structure for each teacher on your team to make sense of *all* students' learning on common assessments by teacher and by team for each essential learning standard.

The intent of the teacher team data dialogue protocols is not to group or level students for initial core instruction, but rather to create a targeted learning plan across the team based on the needs of students to be used in addition to heterogeneous core instruction (NCTM, 2020).

Evidence of Student Learning Protocols

When analyzing evidence of student learning, your teacher team takes action to accelerate learning to grade level or beyond for students who are not yet meeting proficiency, as well as provide extensions for students who are ready to deepen their learning. If your collaborative team stops your common assessment work after creating each assessment, your team fails to fully engage in answering the third and fourth critical questions of a PLC at Work (DuFour et al., 2016)—How will we respond when some students do not learn? and How will we extend the learning for students who are already proficient? Your teacher team will also miss the opportunity to use common assessments for continued learning as part of the formative assessment process.

When analyzing evidence of student learning, you need more than numbers. The calibration routines in chapter 3 (page 47) require your teacher team to consistently interpret the learning shown in student work. It is important to bring that student work to the table (written or annotated on a checklist from a one-on-one or small-group common assessment) when analyzing student learning to see what is revealed in student thinking.

When evaluating student work, your teacher team should do the following.

- Develop and adhere to norms for analyzing student work.
- Score student tasks and sort student work.
- Work collaboratively to identify strengths in student reasoning and common misconceptions students demonstrate in their work.

When your teacher team identifies strengths in student reasoning and uncovers student misconceptions, be sure to focus on students' mathematical reasoning, not their effort. It is easy to blame poor student performance on attendance or failure to complete work. However, to create a targeted response to student learning, your teacher team analyzes the student work produced on the assessment. For example, what are the mathematical strategies students used that were successful or not as successful when showing solution pathways?

Student Work Protocol

The Student Work Protocol (figure 4.1, page 84) is most often used when a teacher team is commonly assessing one essential learning standard with a constructed response (used mostly for common mid-unit assessments). Your teacher team first develops clear expectations for three or four different levels of performance (the assessment may not produce work exceeding the standard, as shown in figure 3.9, page 70).

You can practice using this protocol with one class set of student papers from a designated teacher. However, when using the Student Work Protocol for your full teacher team data analysis and action, each teacher should bring their collection of student work from the assessment. The conversation that results from using the protocol is an example of a team data dialogue.

The Student Work Protocol provides an opportunity for teachers on your collaborative team to see trends in student thinking and make a targeted and specific plan to re-engage students in learning, if needed. Mona's story shares an algebra 1 team's analysis and plan using the data dialogue protocol.

When each teacher on your team brings student work, you and your colleagues take ownership and responsibility for accurately and consistently scoring student work for equitable feedback. Together, you answer the four critical questions by clarifying learning expectations and making a plan to address students who have or have not yet learned the essential learning standard (see chapter 6, page 121).

Figure 4.2 (page 85) shares the directions for the Student Work Protocol. Think about an upcoming assessment and how this protocol could support learning about effective instructional practices and student learning across your teacher team. In the Student Work Protocol, your team analyzes student learning as a progression from below standard to exceeding standard rather than only looking at the percentage of students in each category. This deeper analysis promotes actionable discussions between you and your colleagues, moving past correct versus incorrect student work.

Personal Story **MONA TONCHEFF**

When I used the Student Work Protocol with an algebra 1 team, members noticed students were struggling with using the quadratic formula. As the teams sorted student work, I overheard team members stating they would need to reteach everything about the quadratic formula.

After inspection of the student work during the protocol, the team realized that students actually understood the formula and when to use the formula. However, when using the formula, students were making mistakes with the calculations related to $\sqrt{b^2 - 4ac}$. The teachers were able to create a plan for students to re-engage with the procedural aspect of simplifying expressions involving square roots, squares, and negatives, and they avoided spending an entire day reteaching a concept that students actually understood.

Student Work Protocol

Learning targets or essential learning standards:

1. Analyze the expectation for student work or performance at each level. What are the criteria to assess this work?

Below Standard	Approaching Standard	Meeting Standard	Exceeding Standard

2. Sort student work, and list student names at the appropriate level or the number of students by teacher in each level (with student names identified in the gradebook). Identify what percentage of students are at each level across the team.

Below Standard	Approaching Standard	Meeting Standard	Exceeding Standard
Percent of team:	Percent of team:	Percent of team:	Percent of team:

3. Choose a couple of student samples from each level, and describe the evidence of student learning. Focus on the following four areas. Document the trends of student thinking in each level.
    a. Demonstrates deep conceptual understanding
    b. Shows procedural knowledge of mathematical content
    c. Demonstrates skills and understanding in problem solving
    d. Demonstrates effective communication

Below Standard	Approaching Standard	Meeting Standard	Exceeding Standard

4. Determine a plan for next instructional steps as a team. What is the targeted and specific plan for each group of students to re-engage students in learning (Tier 1 or Tier 2 interventions)? Identify any future changes in instruction on your team unit plan.

Below Standard	Approaching Standard	Meeting Standard	Exceeding Standard

Figure 4.1: Student Work Protocol.

*Visit **go.SolutionTree.com/MathematicsatWork** for a free reproducible version of this figure.*

Protocol Steps (The following four steps take about forty-five minutes to complete.)	Directions
1. Establish the purpose for using the Student Work Protocol.	This protocol will help your team answer the following five questions. 1. What are the differences in student work between the rubric levels *below, approaching, meeting,* and *exceeding* proficiency? 2. What percentages of the students are at each level in individual classes and across the team? 3. What specific misconceptions can you address that may be across all four levels of proficiency? 4. Which instructional strategies are working? Which are not working? 5. Are there re-engagement or enrichment strategies your team can collectively use?
2. Collect the evidence.	Agree on the common assessment to analyze. Team members bring their individual data and student work to the team meeting in preparation for the team data dialogue. *Note: Before using this protocol, teams agree on the scoring and define the four levels of proficiency for the learning targets or essential learning standard.*
3. Analyze the evidence.	If the team has not done so already, members define the four levels of proficiency. Each teacher sorts one class of scored student work. Once sorting of student work is complete, list the names of the students at each level or write the number of students by teacher in each level and then calculate the percentage of students in each level as a team. Choose a couple of student samples from each pile, and analyze and summarize their performance. List observations and share results across the team.
4. Plan action steps based on the results of the analysis.	List two or three action steps based on evidence of learning. Identify content needs and areas for extension. List different instructional strategies to utilize for re-engagement or extension for each group of students.

Figure 4.2: Student Work Protocol instructions.

*Visit **go.SolutionTree.com/MathematicsatWork** for a free reproducible version of this figure.*

TEACHER *Reflection*

Are there specific essential learning standards that students continue to struggle with learning each unit and each year?

How might this Student Work Protocol support your team's exploration of improved strategies for student learning?

When your teacher team analyzes student work, there will be intentional opportunities to identify specific mathematics learning gaps and strengths. A natural result of your teacher team's analyzing student errors and misconceptions (the third step of the Student Work Protocol—analyze the evidence) is the sharing of instructional approaches or strategies that most positively impacted student learning. This, in turn, builds shared knowledge of effective practices and contributes to your teacher collective efficacy.

Essential Learning Standard Analysis Protocol

Another data dialogue protocol your teacher team can use is the Essential Learning Standard Analysis Protocol. This protocol is often used with common assessments addressing more than one essential learning standard (most often common end-of-unit assessments). Your teacher team engages in meaningful conversations around evidence of student thinking aligned to the essential learning standards.

In a PLC at Work, part of your teacher team's collective response involves creating a tiered required response to student learning (see chapter 6, page 121). The Essential Learning Standard Analysis Protocol provides a framework to help you determine the essential learning standards some students still need to learn and those that could be extended based on evidence of student learning. Figure 4.3 shows an example of a grade 4 team's analysis of a common end-of-unit assessment related to fractions from figure 3.5 (page 71).

Sharing results of student learning (or a lack thereof) can be difficult due to a fear of being judged by colleagues or administrators. Keep your conversations focused on strategies successful students used to demonstrate a deep understanding of the essential learning standards. Also identify students who need additional supports or extension. In other words, keep the focus on student learning, and your teacher team insights related to instructional practices that work best will follow.

After you collect your student performance data individually, within a few days of giving the common mathematics assessment, you then bring that collected evidence to your team meeting. Your team can use Google Docs or a similar digital resource to create a form that collects the student learning results across the team. Once your team determines the percentages of students in each level of proficiency (by essential learning standard), the team can review the results and discuss next steps during a team meeting.

When analyzing student work, be sure to bring your student papers or checklists to the table. Student work may need to be referenced or further analyzed by the teacher team for effective interventions. Sarah's story shares one example of a team effectively pausing in order to re-examine student work.

Read the instructions for the Essential Learning Standard Analysis Protocol in figure 4.4 (page 89). Think about an upcoming assessment and how this protocol could be used to support the professional work of your team.

When your grade-level or course-based collaborative team engages in an analysis of student learning, standard by standard, your teacher team develops a better understanding of the expected student proficiency levels for each essential learning standard. Use the following teacher reflection to consider how to best use the protocol presented in figure 4.3 to support the formative mathematics assessment process for your team.

TEACHER *Reflection*

How can your teacher team use the Essential Learning Standard Analysis Protocol to support your team's understanding of student learning for each unit of mathematics?

_____
_____
_____
_____
_____
_____
_____

Personal Story **SARAH SCHUHL**

When I used the Essential Learning Standard Analysis Protocol with a grade 1 team, one teacher had more proficient and advanced students than any other team member; however, when she shared her student work, the evidence of proficient and advanced was not apparent, so the team questioned her. She replied that she knew the students understood the concepts from their work in class—they just didn't show all of their thinking on the test.

We had to stop and talk about how looking at student work requires an understanding that the student thinking shown alone determines proficiency and the effectiveness of instruction, and that teachers need to work with students to understand the expectations in writing their thinking and solutions to tasks.

Essential Learning Standard Analysis Protocol

1. Define each essential standard on the assessment and describe the expectation of proficiency. Write your definitions and descriptions in the following chart.

Expectations of Proficiency	Essential Standard 1	Essential Standard 2	Essential Standard 3	Essential Standard 4
	I can explain why fractions are equivalent and create equivalent fractions. Student creates two equivalent fractions from a given fraction or model and explains why the fractions are equivalent.	I can compare two fractions and explain my thinking. Student compares two fractions using <, >, or = when given a model and without a model and shows an understanding that fractions can only be compared when they reference the same whole.	I can add and subtract fractions and show my thinking. Student adds and subtracts fractions with like denominators, decomposes a fraction more than one way, and shows thinking using models and equations.	I can multiply a fraction by a whole number and explain my thinking. Student multiplies a fraction by a whole number using models and equations to show thinking.

2. Determine the number and percentage of students proficient or advanced on the assessment for each standard by teacher, and then determine the total number of proficient or advanced students within the team. Write the information in the following chart.

	Essential Standard 1		Essential Standard 2		Essential Standard 3		Essential Standard 4		Total Number of Students
	Number	Percent	Number	Percent	Number	Percent	Number	Percent	
Teacher A	20	65%	10	32%	22	71%	28	90%	31
Teacher B	19	68%	12	43%	20	71%	26	93%	28
Teacher C	20	65%	8	29%	28	90%	25	81%	31
Total Team	59	66%	30	33%	70	77%	79	88%	90

3. For each standard, determine which students have unsatisfactory knowledge, which have limited knowledge, which are proficient, and which are advanced by teacher and as a team.

Essential Standard 1

	Beginning	Approaching	Proficient	Advanced	Total Number of Students
Teacher A	2	9	10	10	31
Teacher B	8	1	19	0	28
Teacher C	11	0	16	4	31
Total Team	21	10	45	14	90

Essential Standard 2

	Beginning	Approaching	Proficient	Advanced	Total Number of Students
Teacher A	10	11	2	8	31
Teacher B	10	6	12	0	28
Teacher C	19	4	4	4	31
Total Team	39	21	18	12	90

continued →

Figure 4.3: Essential Learning Standard Analysis Protocol—Grade 4 sample.

Essential Standard 3	Beginning	Approaching	Proficient	Advanced	Total Number of Students
Teacher A	4	5	7	15	31
Teacher B	7	1	19	1	28
Teacher C	3	0	18	10	31
Total Team	14	6	44	26	90

Essential Standard 4	Beginning	Approaching	Proficient	Advanced	Total Number of Students
Teacher A	1	2	20	8	31
Teacher B	0	2	22	4	28
Teacher C	6	0	24	1	31
Total Team	7	4	66	13	90

4. Which essential standards were student strengths? What instructional strategies impacted student thinking?

Our students are doing well with adding and subtracting fractions with common denominators and multiplying fractions by a whole number. Having the students engage in number talks during this unit has really helped students make connections between the models that students create and develop the thought processes needed.

5. In which areas did individual teachers' students struggle? In which areas did our team's students struggle? What is the cause? How will we respond?

Only 66% of our students are proficient with equivalent fractions. Teacher A has been using more manipulatives and will use them with students we define need intervention from all three classrooms.

As a team, students are struggling with comparing two fractions when the denominator is not common. They are confusing the models or are not precise when using a circle to compare fractions. We will create a plan for the students who need more time and support to include a focus on fraction representation using the rectangular model and the applet from NCTM.

6. Which students need additional time and support to learn the standards? What is our plan?

Next week during intervention time, we will use the following schedule:
Monday and Tuesday—Teacher A and support staff will work with the 39 identified students on comparing two fractions. Teachers B and C will use the recipe task to stretch students' understanding of uncommon denominators.
We will also use small-group instruction and centers during the week to do more work with manipulatives. Teacher A is going to bring her manipulatives to share at the next team meeting.

7. Which students need extension or enrichment? What is our plan?

See notes on the use of the recipe task.

Source: Adapted from Kramer & Schuhl, 2017.

Visit go.SolutionTree.com/MathematicsatWork for a free reproducible version of this figure.

Protocol Steps (The following four steps take about forty-five minutes to complete.)	Directions
1. Establish the purpose for using the Essential Learning Standard Analysis Protocol.	This protocol will help your team answer the following seven questions. 1. How many students are beginning, approaching, proficient, or advanced by teacher and as a team? 2. What specific misconceptions can you address that may be across all four levels of proficiency? 3. Which instructional strategies are working? Which are not working? 4. What re-engagement or enrichment strategies can your team collectively use? 5. In which areas did individual teachers' students struggle? In which areas did your team's students struggle? What is the cause? How will you respond? 6. Which students need additional time and support to learn the standards? What is your plan? 7. Which students need extension or enrichment? What is your plan?
2. Collect the evidence.	Agree on the common mathematics unit assessment to analyze. Be sure the questions align to the essential standards prior to compiling the student results. Each teacher documents the number of students in each learning level by essential learning standard and the percentage of students proficient or advanced, ideally before going to the meeting after having scored student papers. The team members will bring student work to the team meeting and will be prepared for the discussion. *Note: Before the team uses this protocol, team members need to reach agreement on the scoring of the mathematics tasks for each essential learning standard and expectations of proficiency for those standards.*
3. Analyze the evidence.	The team determines the number of students across the team in each proficiency level for the essential learning standards as well as the percentage of students proficient or advanced for each essential learning standard across the team. Complete the team reflection questions using examples from student work to identify trends in student learning. *Note: It is critical for all teachers on the teacher team to bring their student work to the meeting for this analysis.*
4. Plan action steps based on the results of the analysis.	List two or three team action steps based on evidence of learning. Identify content needs and areas for enrichment or extension. Create additional team interventions as needed to address the learning needs of students across the team.

Figure 4.4: Essential Learning Standard Analysis Protocol instructions.

*Visit **go.SolutionTree.com/MathematicsatWork** for a free reproducible version of this figure.*

As a result of using the Essential Learning Standard Analysis Protocol, your mathematics team will be able to identify specific students who meet your proficiency expectations by standard and then target those students in need of additional time and support.

Your team can also use the student data to discuss instructional practices that may have been more effective (or not) toward helping students learn essential standards as revealed in student work. From multiple-choice questions, your team might look at commonly chosen distractors that reveal student misconceptions and can inform next instructional steps as a team. This level of data dialogue allows you and your colleagues to create a collective team response to student learning as part of a Tier 2 intervention plan (see chapter 6, page 121).

Your teacher team's collective response will address common student errors revealed when your team analyzes student reasoning on common assessments. Together, you and your colleagues will identify meaningful instructional strategies and mathematical tasks to re-engage students in learning related to an essential learning standard for the mathematics unit. In turn, these actions will result in more equitable student learning experiences by providing identified and targeted additional mathematics learning experiences for all students and all members of your teacher team.

Some of the actions your team can take include the following.

- Use a buffer or flex day in the unit-planning calendar to schedule small-group student-engaged instruction for addressing student misconceptions.
- Determine learning targets to address during an intervention time in the school schedule; identify a new instructional approach and determine which teacher on the team will support students in need of intervention for specific mathematics learning targets.
- Share students during core instruction or intervention time; for example, students who need support with the first essential learning standard go to teacher A, students who need support with the second essential learning standard go to teacher B, and so on.
- Add additional warm-up activities or prompts for two weeks to see if a different strategy supports student proficiency toward a specific mathematics learning standard.

As part of a natural progression of your team's assessment work, you will identify essential learning standards that are more of a challenge for your students to meet proficiency. You will transition to sorting student work based on proficiency levels, analyze the actual student thinking to understand student errors, measure students' current level of mastery, and provide time to improve their understanding and learning for the specific essential mathematics standard (further discussed in chapter 6, page 121).

If the grade 4 example in figure 4.3 (page 87) were your collaborative team's analysis, you might notice that one team member identified more students at the advanced level and one team member had several students at the beginning level. Discussion of the team data might reveal that one instructional strategy impacted student understanding more than another. In fact, only one teacher might have used the strategy. If that is the case, then during a scheduled time for intervention, the teacher who used the more effective strategy might work with all the students across the team who still need to learn the specific essential learning standard.

Looking at student thinking will prompt deeper discussions between you and your colleagues as you seek to positively impact student learning. You will notice your team asking, "Which specific instructional strategies used during our core instruction promoted deeper student understanding, and which actions did not?"

As you and your team gain more experience with protocols for analyzing student work (written or annotated on a checklist), your teacher team should modify and adjust any protocol to meet your specific needs for common assessment analysis and action.

Student Reasoning Protocol

The Student Reasoning Protocol is another structure your teacher team can use to evaluate (1) the effectiveness of an assessment with either a singular mathematical task or a few tasks and (2) the level of thinking, solution pathways, selection of strategies, or problem solving students use to complete the task or tasks. Notice that this protocol involves collecting less assessment evidence, as you and your colleagues focus on the mathematical thought processes for a specific learning target or essential learning standard.

The Student Reasoning Protocol (figure 4.5) is best used with a higher-level-cognitive-demand mathematical task, or set of tasks, that requires students to explicitly describe and justify their reasoning and allows for multiple entry points and solution pathways (often a common mid-unit assessment or a common task for your teacher team to stop and collectively determine how well students are learning). Discussions related to instruction and student learning are not as deep, nor do they lead to as many insights, when lower-level-cognitive-demand mathematical tasks are used for the analysis with the Student Reasoning Protocol.

Figure 4.5 is a Student Reasoning Protocol completed by a grade 2 team of teachers. Notice it starts with generic criteria for each level of proficiency, and then from there, your teacher team determines what is specifically needed based on the task, or set of tasks, used. With your teacher team, read through the sample student work and reflect on the suggested additional interventions and team actions. What additional action steps might you and your colleagues create based on the evidence of student thinking?

Student Reasoning Protocol

Part 1: Completed Rubric for the Three Levels of Student Thinking

Team: Grade 2 team
Learning target: I can solve word problems and show my thinking.
Task: David goes to the park. He sees 15 dogs, 23 squirrels, and some birds. Altogether, he sees 52 dogs, squirrels, and birds at the park. How many birds does David see at the park? Show how you know your answer is correct.

Student Thinking Evidence
Directions: Read the following generic levels of student thinking and determine how they may apply to your team's task or set of tasks.

Above proficient:
- Uses an efficient and effective strategy to solve the problem
- Provides a detailed and clear explanation with mathematical representations they used to expand the solution
- Uses appropriate mathematical vocabulary
- Justifies the solution pathway
- Shows complete understanding of the concepts, skills, and procedures

Proficient:
- Uses an effective strategy to solve the problem, but makes minor errors
- Provides a clear and supported explanation with mathematical representations
- Uses some of the mathematical vocabulary
- Includes some justification of the solution pathway
- Shows substantial understanding of the mathematical concepts, skills, or procedures

In progress:
- Uses strategies that are not appropriate for the problem
- Lacks explanation or clarity of explanation
- Lacks precise mathematical language
- Does not always provide clear justification for the chosen solution pathway
- Shows limited understanding of the mathematical concepts, skills, or procedures

Part 2: Team Agreed-On Rubric for the Three Levels of Student Thinking With Student Identification for Each

Student Thinking Evidence
Directions: With your team, define the expectation of student reasoning in each proficiency level. Sort the student work accordingly across the team and list each student name (or write the total number of students) in each level.

Student Thinking Rubric	Student Names at Each Level
Above proficient: Student solves the task with work to show the answer is correct and checks the work or shows two ways to justify the answer, including birds as units at the end of the answer.	Student 1 Student 2 Student 3

continued →

Figure 4.5: Student Reasoning Protocol—Grade 2 example.

Proficient: Student solves the task correctly with work to support the answer and may not have the units as part of the answer. Student may support their answer in more than one way. However, they may have one minor error in one of the representations.	Student 4 Student 5 Student 6 Student 7 Student 8 Student 9 Student 10 Student 11 Student 12
In progress: Student has correct work with an incorrect answer or a correct answer, but the work does not support the answer. Or Student shows minimal understanding in work and answer. (For example, student might solve 15 + 23 + 52 = ? correctly.)	Student 13 Student 14 Student 15 Student 16 Student 17

Student 3 work	Student 8 work	Student 14 work

Part 3: Plan for a Team Response Using Student Thinking Evidence

Directions: Determine your team response to the student learning data shown in student work.

Student Thinking Evidence	Next Steps for Enrichment or Re-Engagement
Above proficient	Students are using graphic organizers effectively and are precise with the units. Provide challenge tasks during enrichment time with numbers larger than 100. Possibly include problems that require more than two operations. Another option is to give students an answer and have them create a two-step word problem that results in the given answer.
Proficient	Continue to work with students to help them develop additional strategies. Include time during intervention for using open number lines and breaking apart numbers. Create a station activity for reinforcing multiple strategies, and when asking students to present their work under the document camera, create a rubric for what students should include in their work.
In progress	Use base ten blocks and the number line to reinforce the concept of a missing addend. Students need more work on taking the given information in the task and creating a bar model to represent what information is missing. Use open number lines in the upcoming unit to reinforce addition.

*Visit **go.SolutionTree.com/MathematicsatWork** for a free reproducible version of this figure.*

As you analyze the student work in figure 4.5, what do you notice about the levels of student thinking? During their analysis, the grade 2 team members noticed that across all three levels, students were not using open number lines. The team was able to create a plan for teaching open number lines during the upcoming unit.

The instructions for the Student Reasoning Protocol are shown in figure 4.6. Think about an upcoming higher-level-cognitive-demand mathematical task everyone on your teacher team could give aligned to an essential learning standard for which this protocol would grow teacher and student learning.

Protocol Steps (The following four steps take about sixty minutes to complete.)	Directions
1. Establish the purpose for using the Student Reasoning Protocol.	This protocol will help your team answer the following six questions. 1. What solution pathways are students using to solve this task or these tasks? 2. What are the most effective solution pathways? 3. What problem-solving strategies are students using? 4. Which instructional strategies are working, and which should you modify to encourage more productive problem-solving strategies? 5. Which students need additional time and support to develop more meaningful strategies? 6. Which students need extension or enrichment? What is your plan?
2. Collect the evidence.	Agree on the common task or set of tasks to analyze. Be sure to focus on a problem-solving strategy or strategies to evaluate when compiling the student results. Team members should bring a class set of scored student work to the team meeting and be prepared for the discussion. *Note: Before using this protocol, teams agree on the scoring of each essential learning standard and target and define the three levels of proficiency.*
3. Analyze the evidence.	Using the Student Reasoning Protocol, the team will either use the generic rubric with defined levels of thinking (part 1) or define and agree on three levels of student thinking (part 2). Each teacher will sort one class of student work based on the evidence of student thinking. Once the team sorts the work, list the names of students at each level. In part 3, each team member describes the evidence of student thinking and the team makes a collective plan to re-engage students in learning, if needed.
4. Plan action steps based on the results of the analysis.	List two or three action steps based on evidence of learning.

Figure 4.6: Student Reasoning Protocol instructions.

*Visit **go.SolutionTree.com/MathematicsatWork** for a free reproducible version of this figure.*

Read the story from Georgina Rivera (p. 94), in which she explains working with a team of teachers to uncover instructional strategies that were getting in the way of student learning and the team response that followed.

Georgina's story shows how the teacher team learned from looking at student work and made a collective plan to re-engage students in learning how to read, make sense of, and solve word problems. After enacting their plan, more students learned.

When your teacher team engages in careful examination of student learning, you open the door for you and your colleagues to take risks and improve teaching

Personal Story GEORGINA RIVERA

While I was working with a grade 2 team, teachers expressed a concern about how to best teach solving different word problem types. When the team reviewed student work, they noticed students were unsure whether to add or subtract, even though their computation that followed was correct. The team had used key words as a strategy, and it was not working. We brainstormed strategies to implement during their intervention blocks and their daily prior knowledge routines to help students better understand the problem types. As a team, they agreed to do the following.

- Select problems that were of high interest, contextual, and culturally relevant.
- Mix the problem types when presenting them instead of presenting one problem type in isolation.
- Use the three reads protocol. (Students read the word problem three different times, each time with a specific purpose: [1] explain the context of the problem, [2] identify the question being asked, and [3] identify what is known and unknown along with a strategy to solve the word problem.)
- Block out the numbers when presenting the story problems so the students would focus on the context and what the question was asking.
- Have students draw visuals to help them make sense of the problems, including number bonds, part-part-whole diagrams, equations, and pictures.
- Have students explain why they chose an operation and how their answer made sense.

After implementing the agreed-on strategies and routines, the team began to notice a shift in student thinking. Instead of just pulling two numbers from the word problem to randomly add or subtract, students were taking more time to make sense of the problem. They started to ask questions of themselves and their peers: Where am I in this problem? What is happening? What are they asking? All this mathematical reasoning helped students better understand that solving mathematics problems requires them to slow down and make sense of the problem so they can create a solution strategy.

For the team, it was a considerable shift in thinking about word problems because they had all solved word problems in school using key words and taught their students to do the same. They also had never used visual or graphic organizers to make sense of problems and organize thinking. In the end, the team realized that solving story problems was more about sense making and less about trying to help students memorize key words. They also loved the three reads protocol because it integrated literacy and mathematics. In the end, this was a learning experience for both the teachers and students that led to increased student learning!

practices. Your teacher team also ensures more equitable opportunities for students to learn or re-engage in learning, regardless of which mathematics teacher a student is assigned.

Engaging in the protocols throughout this chapter, your teacher team begins to shift from using assessments to determining how many students learned, to examining student work for insights into what students learned and how well they learned it. Together, you and your colleagues start to identify effective instructional strategies and create a collective re-engagement plan for continued student learning, whether in the current unit or the next.

TEACHER *Reflection*

How do you and your team determine the student work needed to demonstrate various levels of proficiency with an essential learning standard or task? What might you need to strengthen as part of your team practices while you analyze student work to determine effective instructional practices and design a collective response to student learning as a team?

_____
_____
_____
_____
_____
_____
_____
_____
_____
_____

In a nutshell, your teacher team learns together as you do the following.

1. Create the common assessment together with scoring agreements defining proficiency (see chapter 2, page 17).
2. Score student work, making sure each teacher on your team is calibrated in their scoring (see chapter 3, page 47).
3. Engage in a meaningful discussion about student learning using one of the protocols shared in this chapter along with student work.
4. Discuss effective instructional practices and whether new practices need to be explored.
5. Identify student learning successes, errors, and misconceptions and develop a team plan to respond to that learning.

TEAM RECOMMENDATION

Engage in Teacher Team Data-Analysis and Action Routines

- Choose a data dialogue protocol to use as a teacher team when examining student learning on a common assessment together.
- Identify trends in student thinking and a specific and targeted team response regardless of the protocol used.
- Analyze the effectiveness of instructional practices teachers across the teacher team used.

Students show their learning on state or provincial assessments and progress-monitoring assessments as well as on common mid-unit and end-of-unit assessments and throughout daily lessons. In their book *Street Data*, Shane Safir and Jamila Dugan (2021) use the term *map data* for your common assessment data. When map data are analyzed by your teacher team in real time, that analysis provides a detailed direction for continued student learning. This becomes an important part of your data-analysis routine as a team.

Student work shown on each common assessment (written or annotated on a checklist) informs the feedback and instruction you and your team members use in the classroom as students demonstrate progress toward each essential mathematics learning standard. The common assessments also inform the feedback and continued instruction students will need after each common mathematics assessment. These assessment actions become your team routines to analyze and take action on student learning data.

Who learns from the common assessment data is not limited to the teachers or support members of your team. Formative learning includes the students as assessors of their own mathematics learning as well. In the next chapter, you will explore student actions as part of the formative assessment process related to the common unit-by-unit assessments.

Visit **go.SolutionTree.com/MathematicsatWork** for free reproducible versions of tools and protocols that appear in this book, as well as additional online only materials.

CHAPTER 5

Student Actions in the Formative Assessment Process

Assessment is a process that should help students become better judges of their own work, assist them in recognizing high-quality work when they produce it, and support them in using evidence to advance their own learning.

—NCTM, Principles to Actions

What happens across the members of your collaborative team when you return a common mathematics assessment to your students? What do you expect students to do with the feedback you provide? How does your teacher team ensure the common assessment instrument is a tool that enhances student learning?

It is only through the nature of the expected student response to your feedback during and at the end of the unit that your common assessments become formative for students (Chappuis & Stiggins, 2020; Moss & Brookhart, 2019).

As shown in the common assessment formative process rubric in figure 1.2 (page 12), common assessments are always aligned to the essential learning standards, and written using the criteria for high-quality common assessments (figure 2.1, page 18). The teachers on your team have also calibrated their scoring for consistent feedback to students. Just like you have data-analysis and action routines, so do your students. With clarity of feedback across your teacher team and student involvement with the essential learning standards throughout the unit, your students can accurately reflect on their learning of essential learning standards and set goals with a plan for continued learning.

In this chapter, student actions taken after a common end-of-unit assessment are addressed first, followed by student reflection and action throughout the unit and on common mid-unit assessments. Any tools used for student reflection and action should be used consistently across your teacher team.

You may also want to consider having common student reflection tools as a mathematics department or vertical grade-level teams. Students learn how to interact with and learn about their learning using the tool without a need to reteach how to learn from the tool year to year.

Student Action After the Common End-of-Unit Assessment

When you return student work with your feedback on the common end-of-unit assessment, your students need to self-assess if they are meeting the proficiency targets set for each essential learning standard for that unit. Your FAST (fair, accurate, specific, and timely) feedback allows students, with your support and the support of their peers, to engage in activities that lead to meeting the proficiency expectations for the essential learning standards of the current unit, despite moving on and into the next mathematics unit for learning.

How will your teacher team provide quality feedback to students and then require students to take action on your feedback as part of their continuous learning? The feedback is more than a number or grade. Your team creates conditions for students to be able to determine specifically what they have learned or not learned, yet.

Students can articulate what they have learned and what they have not yet learned as early as preK or kindergarten through high school. Students can reference their learning using essential learning standards if you share those standards at the beginning of a unit, throughout the unit within instruction, as

> **TEACHER** *Reflection*
>
> How do your students make sense of their end-of-unit assessment results based on feedback received for each essential learning standard? For which standards have they demonstrated evidence of learning? Which have they not learned yet? What will you expect them to do if they are not yet proficient with certain standards?
>
> _____
> _____
> _____
> _____
> _____
> _____
> _____
> _____
> _____
> _____
> _____
> _____

an organizational driver of homework assignments, and during reflection and planning after a common end-of-unit assessment. Tim Brown and William M. Ferriter (2021) share that your team's intentional focus on essential learning standards throughout a unit helps students interpret their learning from the common assessments more accurately.

First graders might say, "I can compare numbers, but I can't write my numbers to 120 by myself yet." Similarly, a high school student might say, "I can determine the end behavior of a polynomial function, but I still need to learn how to determine the zeros for those polynomial functions."

These types of statements provide hope to students because students realize they have learned some concepts and skills, and therefore only need to focus on a limited number for continued focus and proficiency. A student is not "bad" at mathematics but rather has a few essential learning standards yet to learn well and needs additional practice and time. If a student still needs to learn every essential learning standard from the common mathematics assessment, being able to name what needs to be learned makes that learning doable, but saying, "I can't do chapter 4," leaves the student with no place to start their learning.

The story from Bill Barnes reveals how student goal setting and action in response to your assessment feedback might have a few road bumps.

> *Personal Story* **BILL BARNES**
>
> When Elizabeth Kunstman was a mathematics team leader at Edison Middle School in Green Bay, Wisconsin, her team found the commitment to high-quality student reflection and goal setting to be incredibly time-consuming and cumbersome. Providing assessment feedback to students, monitoring student analysis of strong- and weak-performing standards on the assessment, and ensuring student response to re-engage in learning required a massive paper trail for teachers.
>
> Elizabeth's team members figured out an efficient way to use technology to help them with this aspect of their formative learning work. Team members each asked their students to utilize the Google Sheets tool to record returned assessment scores by standard. This process enabled students to analyze performance on each essential learning standard and set improvement goals for engagement and action. As an additional benefit, Elizabeth's team also received a fully populated spreadsheet of student data by standard to use for team analysis of overall student performance by standard for the unit.

Elizabeth Kunstman's teacher team pivoted and used Google Sheets to record assessment scores and engage students in the goal-setting process. How can you and your colleagues take action after every common end-of-unit mathematics assessment? Consider the following two teacher team actions.

1. Create a student goal-setting reflection process for students to identify errors and misconceptions by essential learnings tandard and use the assessment results to form a plan of action.

2. Create a process for students to act on their plan (and allow them to improve their score for each essential learning standard on the end-of-unit assessment).

Merely assigning a summative grade to an assessment is the least useful strategy to motivate students to further their understanding (Kanold, Briars, Asturias, Foster, & Gale, 2013). Wiliam (2011) asserts, "As soon as students get a grade [on a test], the learning stops" (p. 123).

Learning should not stop when your collaborative team provides meaningful assessment feedback to students; you might be surprised to know that their score or grade is the least important aspect of the assessment. Your students should be required to take action on their feedback from the unit assessment and use it as a formative learning opportunity. Do not allow the assessment to be a one-and-done approach. Your feedback on students' solutions allows your students to reflect on their strengths and challenges from the common assessment and allows the assessment to be used as part of each student's learning story. This type of student action routine to the common assessment feedback turns your common assessments into a *process* for the student rather than a singular event.

When you pass the test back in class, give students time to self-assess their results and make a plan of action to retool their learning. Wiliam (2018) describes that effective feedback should:

> cause thinking by creating desirable difficulties. Feedback should be focused; it should relate to the learning goals that have been shared with the students; and it should be more work for the recipient than the donor. Indeed, the whole purpose of feedback should be to increase the extent to which students are owners of their own learning. (p. 153)

To ensure student ownership of learning with continued action, your team's feedback to students becomes much more than a score. Figure 5.1 (page 100) is a sample student self-assessment form for the grade 4 unit assessment on fractions in figure 3.5 (page 54). This self-assessment tool also aims to be a communication tool sent home for parents or guardians to sign as an acknowledgment of their student's progress in learning mathematics. Notice how much more information this reflection provides than an overall percentage or score on the assessment. This reflection tool takes advantage of the intentional team design of organizing the common mathematics assessment by essential learning standard.

TEACHER *Reflection*

How do your students learn from their errors or misconceptions on end-of-unit assessments? How do they equate that learning to an understanding that they are learning an essential learning standard, rather than just simply learning how to do a specific task?

_____
_____
_____
_____
_____
_____
_____
_____

Figure 5.2 (page 100) highlights another sample protocol for student self-reflection and use after the teacher grades and returns the sample high school algebra 1 / integrated mathematics I unit test (figure 3.6, page 58). Your teacher team may decide to use percentages as the basis for identifying what students have learned or not learned yet, instead of points from the scoring guide, as shown in figure 5.2, or your team could use proficiency rubric scores.

Grade 4 Fractions Unit
Student Reflection: Unit Assessment—What Have I Learned? What Have I Not Learned Yet?

Name: _____ Date: _____

Essential Learning Standards	Test Questions	Score	How I Did (Circle one.)	
1. I can explain why fractions are equivalent and create equivalent fractions.	1–4	___ out of 8	I got it! 6–8	I'm still learning it. 1–5
2. I can compare two fractions and explain my thinking.	5–9	___ out of 12	I got it! 9–12	I'm still learning it. 1–8
3. I can add and subtract fractions and show my thinking.	10–14	___ out of 15	I got it! 11–15	I'm still learning it. 1–10
4. I can multiply a fraction by a whole number and explain my thinking.	15–17	___ out of 7	I got it! 5–7	I'm still learning it. 1–4
5. I can solve word problems involving fractions.	18–20	___ out of 6	I got it! 4–6	I'm still learning it. 1–3
Learning standards I know and can do:		**Learning standards I am still learning:**		

Figure 5.1: Sample student end-of-unit self-assessment—Grade 4 fractions unit.

Name: _____ Date: _____

Algebra 1 / Integrated Mathematics I Linear Functions Unit
Use the following to reflect on the questions: What have I learned? What have I not learned yet?

Essential Learning Standards	Test Questions	Score	Learned or Not Yet?	
1. I can graph linear functions and interpret their key features.	1–5	___ out of 13	Learned 10–13	Not yet 0–9
2. I can calculate and explain the average rate of change (slope) of a linear function.	6–8	___ out of 6	Learned 5–6	Not yet 0–4
3. I can recognize situations with a constant rate of change and interpret the parameters of the function related to the context.	9–12	___ out of 11	Learned 8–11	Not yet 0–7
4. I can construct linear functions given a graph, description, or two ordered pairs.	13–17	___ out of 18	Learned 13–18	Not yet 0–12
5. I can graph piecewise functions, including step and absolute value functions.	18–21	___ out of 16	Learned 12–16	Not yet 0–11

Essential learning standards I learned:

Essential learning standards I am still learning and why (for example, I made a small mistake, I don't understand, or I need more practice):

My plan to learn each essential learning standard I have not yet learned (include specific tasks assigned by the teacher, use of the tutor center, WIN time, or other approved learning activity):

My goal for the next assessment:

Figure 5.2: Sample student end-of-unit self-assessment—High school linear functions unit.

There are many options with the focus not so much on the score as on use of the score as evidence of student proficiency and learning. Your students will learn quickly how to focus their future time and effort as they analyze and learn from the feedback on their solutions to each task on their assessments.

The intent of students' using the self-reflection form in class when you pass back the end-of-unit mathematics assessment is to help them self-assess their performance and build responsibility for their own learning. From the reflection, students can set learning actions around those concepts and skills not yet learned. A student self-assessment and goal-setting protocol works when students have interacted with the essential learning standards in their daily lessons so they more deeply understand the language on the assessment and assessment reflection tool. (See *Mathematics Instruction and Tasks in a PLC at Work, Second Edition*.)

Your teacher team can also determine how all students will work to learn those standards during the next unit or through the Tier 2 interventions discussed in chapter 6 (page 121).

Note that figures 5.1 and 5.2 show student self-reflection tools that include an analysis of both what students do know and what they do not yet know. This is intentional and important.

One critical factor known to promote access and equity is the use of teaching practices that build on the strengths of students (Fernandes, Crespo, & Civil, 2017). As such, you and your colleagues need to be careful to avoid a deficit orientation with respect to your students' knowledge and skills.

Too often, teachers on a collaborative team focus almost exclusively on what students do not know. However, to support students in recognizing they are mathematically capable, you must also acknowledge what students do know (Jilk & Erickson, 2017). When you acknowledge their strengths, students have the opportunity to develop positive identities of themselves as learners of mathematics (Aguirre, Mayfield-Ingram, & Martin, 2013).

Collaboration with peers and ownership of learning pathways engage students more deeply in the learning process and provide evidence for each student that effective reflection, effort, and subsequent action is the route to mathematical understanding and success.

Formative reflection is not an easy process for students to learn in the beginning. By connecting the primary purpose of your unit-by-unit assessments as less about a grade, and more about progress toward learning, you facilitate an eventual shift in how students prepare for and embrace their assessments as learning opportunities.

Accurate self-reflection by students that allows them to learn from their assessments must often be taught preK–12, especially if students have never before been asked to reflect. Sarah's story illustrates the need to teach meaningful and effective self-reflection.

Personal Story SARAH SCHUHL

I learned quickly that when providing more focused student reflections with error analysis, students viewed every missed question on an assessment as a simple mistake and did not always recognize conceptual errors in reasoning. In fact, after designing a complex reflection sheet and distributing it to my students in calculus and algebra 1 on their respective assessments, I realized that students could not accurately learn from the reflection sheet because they did not understand *how* to reflect in mathematics.

As such, I had to step back and teach self-reflection through sharing samples of student work on the unit assessment, and ask my students to identify the strengths and areas to improve from those anchor papers. Once students were able to see the benefits of exploring mistakes and they were better able to identify strong and weak work, they began to also better self-reflect and learn from their thinking on their assessments.

Consider how you and your colleagues will use a systematic process after each common end-of-unit mathematics assessment to help students learn a process for analyzing their work and interpreting your feedback. While it will take time to teach students this essential reflection skill, over time, students learn to quickly and accurately identify what they have learned and not learned yet, as well as understand the effort required to continue learning due to any errors made on the assessments' mathematical tasks.

At every grade level from preK through high school, students need to learn how to accurately self-reflect on their progress toward learning essential standards and set goals. And, at every grade level, students can do this work independently, in groups, or with a teacher and learn from the results.

At the DuFour Award–winning Mason Crest Elementary School in Annandale, Virginia, kindergarten students have essential learning standards on strips of paper attached to a large ring. One paper, for example, might show the essential learning standard, *I can count to 100*. Underneath the essential learning standard are benchmarks such as 1–10, 11–20, 21–50, and 51–100, each written below a star. As students meet each benchmark on the way to becoming proficient with the essential learning standard, they hole-punch or color the corresponding star. Over the course of the year, students record their continued learning for each essential learning standard (J. Deinhart, personal communication, June 1, 2017).

Figure 5.3 shows an example adapted from Mason Crest of a completed kindergarten reflection tool that can be used after rolling assessments (see figure 3.4, page 53) or after designated times as a common end-of-unit assessment experience. Students hole-punch or color the squares as they continue to learn. The right side of the tool shows a possible extension for kindergartners. There is one reflection card for one essential learning standard. Students may have several learning cards on a ring, one per essential learning standard.

Your teacher team may also decide to use a more complex data-tracking chart like the one in figure 5.4, which the algebra 1 team in the Norwalk Community School District of Iowa used. Students record their current proficiency on each essential learning standard throughout a semester. This kind of recording form can be part of a data notebook and something you use for student-led conferences along with the common assessments and self-reflection forms. Although the recording action presents an overview of their learning, your students need to still learn from their feedback through an analysis and reflection process.

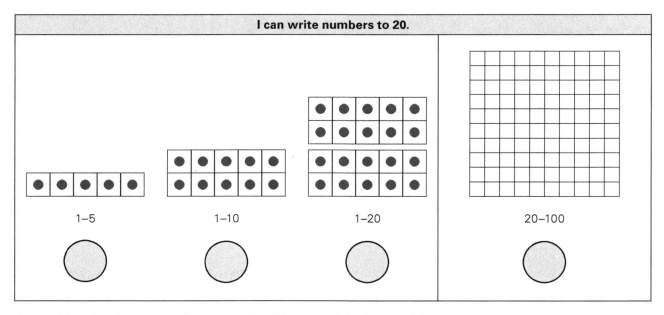

Source: *Adapted with permission from Mason Crest Elementary School, Annandale, Virginia.*

Figure 5.3: Example kindergarten self-assessment.

Student Actions in the Formative Assessment Process | 103

Algebra 1 First Semester Proficiency Graph

Name: _____ Period: _____

Scale		Learning Target	Category
4.0 – 0.0		★ Write Equations of Lines Assessment	Write Equations of Lines
		Lines of Best Fit CFU	
		Parallel and Perpendicular Lines CFU	
		Write Equations of Lines CFU	
		★ Graph Equations Assessment	Graph Equations
		Graph Equations in Standard Form CFU	
		Graph Equations in $y = mx + b$ CFU	
		★ Slope Assessment	Slope
		Slope (Rate of Change) CFU	
		★ Interpret Functions Assessment	Interpret Functions
		Arithmetic Sequences CFU	
		Function Notation CFU	
		Functions, Domain, and Range CFU	
		★ Create and Graph Functions Assessment	Create and Graph Functions
		Write and Graph Functions CFU	
		Write and Graph Function Patterns CFU	
		★ Solve Inequalities Assessment	Inequalities
		Absolute Value Equations and Inequalities CFU	
		Compound Inequalities CFU	
		Solve Multistep Inequalities CFU	
		★ Solve Proportions Assessment	Equations
		Solve Proportions CFU	
		★ Solve Equations Assessment	
		Solve Equations CFU (2.2–2.5)	
		Solve Variables on Both Sides CFU	
		Solve Multistep Equations CFU*	

Proficient line at 3.0

Learning Targets

*CFU = Check for understanding

Color key: ☐ First attempt ☐ Second attempt ☐ Third attempt

continued →

Figure 5.4: Student progress-tracking chart from algebra 1 in Norwalk Community School District.

Algebra 1 Second Semester Proficiency Graph

Name: _____   Period: _____

Learning Targets	Category
★ Solve Quadratics Assessment	Solve Quadratics
Solve Quadratics CFU	
Solve Quadratics CFU	
★ Graphing Quadratics Assessment	Graphing Quadratics
Graphing Quadratics CFU	
★ Factoring Polynomials Assessment	Factoring Polynomials
Factoring Polynomials a > 1, Grouping CFU	
Factoring Polynomials a = 1 CFU	
★ Operations of Polynomials Assessment	Operations of Polynomials
Multiply Polynomials CFU	
Add and Subtract Polynomials CFU	
★ Exponential Functions Assessment	Exponential Functions
Geometric Sequences CFU	
Exponential Growth and Decay CFU	
★ Exponent Properties Assessment	Exponent Properties
Rational Exponents and Radicals CFU	
Exponent Properties CFU	
★ Systems of Linear Inequalities Assessment	Systems of Linear Inequalities
Systems of Linear Inequalities CFU	
★ Solve Systems of Equations Assessment	Systems of Equations
Solve Systems by Elimination CFU	
Solve Systems by Substitution CFU	
Solve Systems by Graphing CFU	

Y-axis scale: 4.0, 3.5, 3.0 (Proficient), 2.5, 2.0, 1.5, 1.0, 0.5, 0.0

Color key: ☐ First attempt   ☐ Second attempt   ☐ Third attempt

Source: © 2017 by Norwalk Community School District. Used with permission.

With your collaborative team focused on student learning, your students begin to understand their end-of-unit reflection will in some way require them to re-engage in learning standards that fall into the *not yet* category as well as celebrate what has been learned.

You and your colleagues determine how students will demonstrate their learning of the mathematics standards when every unit assessment is used as part of a more formative process. You will work to ensure students learn the essential learning standards to grade or course level with continued learning experiences from the feedback, if needed. The formative process requires a team response and a student response to feedback. It also affects your team's grading decisions, discussed in more detail in *Mathematics Homework and Grading in a PLC at Work* (Kanold, Barnes, et al., 2018).

An example of student reflection and learning with an opportunity to retake an assessment and demonstrate additional learning is shown in figure 5.5 (page 106). This is a sample from Oak View Middle School in Andover, Minnesota. Students reflect on their learning using evidence from an assessment and then use the FAST feedback idea to commit to a learning plan (contract) that includes correcting assessment tasks, often collaboratively.

Alternatively, some teacher teams will ask students to sign a contract to retake an end-of-unit assessment that requires the students to articulate which learning targets they learned and which they still need to learn with an accompanying learning plan and agreed-on date by which the learning will occur and the test will be taken.

The intent of having students analyze their performance on the common end-of-unit assessment is to help each student build responsibility for their own learning. While each student takes ownership for their individual progress toward each of the essential learning standards, students may still work together to meet those standards. Students can work together when your teacher team provides equitable feedback to students, regardless of which teacher they are assigned, on your common end-of-unit assessments. Together students can revise and practice tasks and give feedback to one another as they re-engage in learning any essential standards not yet learned.

Student-to-student feedback also is a vital component of the entire learning process (see *Mathematics Instruction and Tasks in a PLC at Work, Second Edition*), and goal setting at the end of each unit is part of student collaboration. As students compare and contrast their work with one another, they see what they learned well and what they still need to learn.

TEAM RECOMMENDATION

Require Student Action on Feedback From the Common End-of-Unit Assessment

- Develop a team systematic form for students to use as a self-reflection tool after each end-of-unit assessment. Be sure students identify what they have learned as well as what they have not learned yet.
- Teach students to analyze their work and accurately self-reflect using the assessment feedback.
- Require students to collaborate, take action, and learn from their mistakes or errors.

TEACHER *Reflection*

How do students track their learning progress over time or after each end-of-unit assessment for your grade level or course?

What are some ideas you are considering to strengthen student reflection and action after each end-of-unit assessment, which you and your colleagues can employ?

Expression Test Reflection

Learning Targets for Grade 8, Unit 4, Part A:

1. I can use the distributive property and combine like terms to simplify algebraic expressions.

2. I can evaluate algebraic expressions.

3. I can translate from a real-world situation to an algebraic expression.

4. I can use formulas with multiple variables.

Retake Option

If a learning target score on your assessment is below 70 percent, you are **required** to retake the assessment to improve your score on the learning target. You must work with another student, a tutor, or your teacher to complete the following problems listed for the learning targets for which you scored below 70 percent.
This may mean you retake all four learning targets or just one of them. Turn in the assessment problems when you come to your scheduled retake time.

Learning target 1: Ask your teacher for extra practice and submit by _____ date.

Learning target 2: Ask your teacher for extra practice and submit by _____ date.

Learning target 3: Pages 1–9, 12, and 20

Learning target 4: Pages 1–8, 304

Learning Contract

In order to retake parts of this test, you must complete the following items.

☐ Complete and staple the unit 4, part A test corrections to the original test (other side).

☐ Complete and turn in all practice from during this unit.

☐ Complete and turn in the study guide for the test.

☐ Complete, grade, and turn in retake practice problems (by standard) for the learning targets for retake.

What other actions will you take to learn each standard from this expressions unit assessment?

_____   _____
Student Signature   Date

Name: _____

Test Correction Form

Attach original test to this form.

Number	Reason I Got This Wrong	Correct Answer	Reason My New Answer Is Correct

Source: © 2014 by Oak View Middle School. Used with permission.

Figure 5.5: Example grade 8 student contract and test correction form.

Collaborating with peers and taking ownership of their learning engages students more deeply in the learning process and provides evidence for each student that effective effort leads to mathematical understanding and success.

Not only does your teacher team engage in student reflection and action routines after each common end-of-unit assessment, but students also engage in reflection and action routines after each common mid-unit assessment.

Student Action After Common Mid-Unit Assessments

In addition to common end-of-unit assessments, students also learn and benefit from their shorter common mid-unit assessments provided during a mathematics unit. Since each assessment closely aligns to one or two essential learning standards, students can use your assessment feedback to determine whether they have learned the standard yet before the final end-of-unit assessment.

Students need to take action on and ownership of the assessment feedback they receive during the unit just as they do after the end-of-unit assessment. They will then be able to engage in mathematics interventions (and extensions) before the unit ends as you help them respond to PLC critical questions 3 and 4 (DuFour et al., 2016): How will we respond when some students do not learn (in this case, during the unit)? and How will we extend the learning for students who are already proficient? Thus, learning during the unit becomes more formative.

NCTM's (2014) *Principles to Actions* states:

> At the center of the assessment process is the student. An important goal of assessment should be to make students effective self-assessors, teaching them how to recognize the strengths and weaknesses of past performance and use them to improve their future work. (p. 95)

Think about who is working harder to accelerate student learning to grade level—you or your students? How can you work together with students to encourage and increase their ownership of learning mathematics?

Your students can participate in their own learning in many ways, which allows them to track their own progress and set their own goals. Two strategies include student trackers and self-regulatory feedback. Additional information about how students can reflect on their proficiency on the essential learning standards and daily learning targets is included in your daily lesson design as described in *Mathematics Instruction and Tasks in a PLC at Work, Second Edition*.

Student Trackers

Ridgeview Middle School in Visalia, California, uses a tracker sheet to build a routine way of having students document their learning toward the essential standards in each unit (see figure 5.6) based on feedback it receives or generates during the unit. Students receive the tracker on cardstock at the start of the unit. At the top of the tracker, essential learning standards appear as student-friendly learning targets. The teachers on the team reference the learning targets throughout the entire unit, identify them for each common homework assignment (for more on this idea, see *Mathematics Homework and Grading in a PLC at Work*), and write them on each common assessment within the unit as well as at the end of the unit for each associated group of questions.

The teacher team has created additional learning opportunities for students, which students choose from on the second side of the tracker while they track their progress. (This is part of the team's Tier 2 mathematics intervention program based on their assessment performance; see chapter 6, page 121.)

TEACHER *Reflection*

How are students able to reflect during a unit on their progress toward meeting an essential learning standard?

When students recognize that they still need to learn essential learning standards during a unit, what structures are in place for them to re-engage in that learning?

Name: _____ Period: _____ Date: _____

Student Tracker Unit 3: Expressions, Equations, and Inequalities

Learning targets:

1. I can simplify algebraic expressions. | Starting... | Getting there... | Got it! |

2. I can write and solve an equation from a word problem. | Starting... | Getting there... | Got it! |

3. I can write and solve an inequality from a word problem. | Starting... | Getting there... | Got it! |

4. I can graph the solution to an inequality and explain how the solution answers the question. | Starting... | Getting there... | Got it! |

Note: Teacher will use one or two warm-up questions that relate to each of the preceding learning targets to show student proficiency. Students must complete the following assignments to prepare for the warm-up questions.

Common Homework Assignments

Target Number	Lesson	Page Number	Problem Numbers	Assignment Status (Complete or Incomplete)

Vocabulary

Term	Description	Example
1. Expression versus equation		
2. Coefficient		
3. Constant		
4. Distributive property		
5. Inequality		
6. Solution to an equation versus an inequality		

Figure 5.6: Grade 7 student tracker—Self-assessment and action plan.

continued →

Test or Quiz Name	Testing Date	Original Test Score (Original score you earned when you took the test)	Correct and Reflect Score (3—Completed correctly 2—Incomplete and needs corrections 1—Never turned in)	Retake Score (Required if your score is less than 2.5)	Gradebook Score for This Test (Most recent score on the original test or the retake)

Check one:
☐ I am maintaining a proficient level for this class (3.0).   ☐ I need to put forth more effort in order to be proficient (3.0).

Action steps for improvement: Place a check mark in the shaded boxes under the tasks you will do to relearn essential standards, as needed.

Test or Quiz Name	Morning (a.m.) Tutoring With Teacher (Available from 7:45 to 8:15)	Afternoon (p.m.) Tutoring With Teacher (Available from 3:15 to 3:45)	Tutoring With a Peer (Before school, at lunch, or after school)	Completed Unit Study Guide	Homework Status (I have completed all my homework for this test.)	Correct and Reflect (I can do this on my own and finish before future due dates.)	After-School Intervention (I need to sign up for this every weekend.)

How Can I Learn the Target and Raise My Grade?

Successful mathematics students:
- Come to class prepared (organized binder with completed homework, pencils, calculator, textbook, and paper) and are ready to learn every day
- Ask for help before they take an assessment, so there are no surprises
- Ask questions in class, do their very best on every assignment, and know where they still need extra practice
- Complete the correct and reflect in a timely manner if they earn a failing grade on any test (Retakes are available once the correct and reflect is done correctly.)

What do you still need to work on in order to be successful this year?

Everyone can be successful!

Student signature: _____   Parent or guardian signature: _____

Source: © 2016 by Ridgeview Middle School. Used with permission.

Students continue monitoring their progress using the shading bars to the right of each learning target, and shade (or erase) their learning progress based on evidence generated from warm-ups, classroom tasks, homework, exit slips, and common checks for understanding used during the unit.

Because students are clear about what they should be learning, they are able to articulate the standards they need help with as students become part of the intervention process in a productive way. Students work to learn the content with other students in the same course who may have different teachers, expanding their resources for feedback. Additionally, the students become more aware of which mathematics standards they most need to study and learn prior to the common end-of-unit assessment.

TEACHER *Reflection*

After you review the student tracker in figure 5.6, think about how your teacher team could integrate a similar system of student ownership for the work of your mathematics class during the unit.

_____
_____
_____
_____
_____
_____
_____
_____
_____
_____
_____
_____

At elementary grade levels, student trackers use the essential learning standards written as student-friendly targets and may also have common homework (intermediate grades) or a more detailed proficiency scale description with sample problems shown, so students better understand the meaning of each scale.

A kindergarten team, for example, used take-home folders for students. The teachers on the team agreed to two mathematics and two English language arts targets for a unit of time. Each teacher glued four Velcro strips to the inside of student folders (right side), and each Velcro strip had a Velcro circle for students to move as they learned the target. On the left side of the folders, the teachers inserted typed learning targets so they could be changed as needed. The added benefit to this practice is parent or guardian knowledge about the most important essential standards students are working on.

Figure 5.7 (page 112) shows an example of a third-grade tracker with a proficiency scale. Students check their progress as they move from one level to the next for each learning target and also stop at key points during the unit to identify their actions to continue learning. A teacher may create anchor charts for each learning target to show students a model for each essential learning standard.

Your teacher team can modify any sample student trackers you use to build routines around student self-assessment and action during a unit. You and your colleagues should *require* students to use trackers and re-engage in learning. Due to your robust response when students are not learning, and your expectations for them to track their own progress, you signal to students that learning the essential standards for mathematics is of utmost importance and their top priority in your class.

Figure 5.8 (page 113) shares an example that clarifies success criteria for the essential learning standards students are learning. It also shows students how to connect their common formative assessments (CFAs) that occur mid-unit to the end-of-unit assessment (topic test).

In the primary grades, students might use the same reflection tool for a common mid-unit assessment and an end-of-unit assessment or use it after each attempt at demonstrating learning on rolling assessments. Figure 5.9 (page 114) shares a first-grade example for which students color stars to show their learning progress once their teacher clarifies the success criteria for learning by using a proficiency rubric.

Name: _____ Date: _____

Student Tracker for Grade 3: Perimeter and Area Unit

In this unit, we are going to learn about area and perimeter. The learning targets with proficiency scales are below. Check each level you can do to show what you have learned so far and what you have not learned yet.

I can find the perimeter of polygons.

		This means I can:	What I Can Do	My Plan to Learn
	4	Create two different polygons that have the same perimeter, and explain how I made them.		
	3	Find the perimeter of a polygon if I know the side lengths or if I have to figure out the side lengths myself.		
	2	Find the perimeter of a polygon when I am shown all the side lengths.		
	1	Find the perimeter of a polygon with help.		

I can find the area of a rectangle.

		This means I can:	What I Can Do	My Plan to Learn
	4	Find a missing side length on a rectangle if I know its area, and find two or more different rectangles that have the same area.		
	3	Find the area of a rectangle using a formula and explain why the formula works.		
	2	Find the area of a rectangle by making an array and counting the unit squares.		
	1	Find the area of a rectangle by counting unit squares.		

I can find the area of a complex shape made of rectangles.

		This means I can:	What I Can Do	My Plan to Learn
	4	Create two different complex shapes made of rectangles that have the same area, and show all the side lengths for each.		
	3	Find the area of a shape made of rectangles if I know all the side lengths I need or if I have to find some of them first.		
	2	Find the area of a complex shape made of rectangles when I am shown all the side lengths I need.		
	1	Find the area of a complex shape made of rectangles when I am shown all the side lengths I need and have some help.		

Figure 5.7: Grade 3 student tracker—Self-assessment and action plan.

Name: _____

My Goal-Setting Sheet: Multiplying Decimals

Learning Target	Essential Learning and Success Criteria			
	Keep Trying	**Starting**	**Getting There**	**Got It!**
1. Multiply decimals by a power of ten.	I am still learning to multiply by powers of ten.	I can multiply by powers of ten, but cannot explain why.	I can multiply by tens or hundreds, but not both.	I can explain the patterns in the placement of the decimal point when a decimal is multiplied by a power of ten.
2. Multiply decimal numbers.	I am still learning how to use models and drawings to multiply decimals.	I can start the model, but can't complete it.	I can multiply but might have one error in either my model or my computation.	I can accurately multiply decimals using concrete models, drawings, or strategies.
In the beginning I was at level _____. My goal at the end of this unit is to be at level _____.				

	CFA 1	CFA 2	CFA 3	Topic Test
Date				
Directions: Circle one level of proficiency for each assessment.	4—Got it! 3—Getting there 2—Starting 1—Keep trying	4—Got it! 3—Getting there 2—Starting 1—Keep trying	4—Got it! 3—Getting there 2—Starting 1—Keep trying	4—Got it! 3—Getting there 2—Starting 1—Keep trying
Reflection I met level _____ because _____. What will I do to improve my understanding? (Use vocabulary from the success criteria to explain.)				

Graph Your Progress

4				
3				
2				
1				
	Date			
	CFA 1	CFA 2	CFA 3	Topic Test

Figure 5.8: Grade 5 student tracker—Self-assessment and action plan.

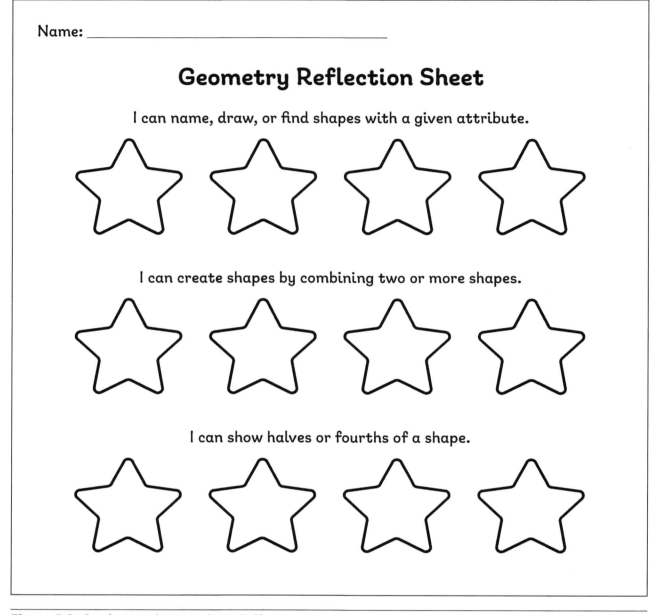

Figure 5.9: Grade 1 student tracker—Self-assessment and action plan.

One final example at the high school level includes students' identifying the percentage earned by essential learning standards on each common mid-unit assessment or quiz as well as on the end-of-unit assessment (see figure 5.10, page 115). Students can determine if there is growth and identify their level of mastery after the unit assessment. After each quiz, students identify, if needed, how they will re-engage in learning the essential learning standard and identify time to take a recovery quiz prior to the end of the unit. The options for continued learning require the team to have a systematic Tier 2 response (discussed in the next chapter).

Once students reflect and take action on their learning, you and your teacher team may want to ask some additional reflection questions to have students think about the effectiveness of their learning plan. Some sample questions follow.

- What parts of your plan worked well? Why?
- What parts of your plan didn't work well? Why?
- What might you do differently in the future?

If common assessments are to be used formatively and be part of a student's learning story, consider as a team how to require students to learn from the feedback given.

Name: _____   Date: _____

Student Tracker for High School: Geometry, Right Triangles, Unit 5

For each essential learning standard, record your points earned on the corresponding points quiz, and then decide how well you understand the learning standards at this time. For the final assessment, record how many points you earned for each essential learning standard, and then determine what kinds of mistakes you made and your target level of mastery for each learning standard.

For each learning target you are approaching, create an action plan of how you will learn the standards, and be prepared to take the recovery quiz (your second attempt at demonstrating mastery). Please remember to also write times and dates. The more specific you are about your plan, the more likely you are to stick to it. "Whenever" is not specific!

		Point Quiz			Unit Test				
	Checkpoint quizzes are given to monitor learning and to address misconceptions with tutoring before the unit test.	Right Triangles Readiness Check How well do I know it?			Level of Accuracy Why did I not earn full credit?			Level of Mastery (in percentages) Exceed (90–100) Proficient (70–89)	
Standards	Learning Target (I can. . .)	Points Earned	Percentage	Tutoring (Circle yes or no.)	Questions on Test	Points Earned	Percentage	Approaching (50–69)	Some Evidence (25–49)	No Evidence (0–24)
G-SRT.6	Explain the definitions for trigonometric ratios for acute angles in right triangles using similarity.	___ out of 8		Yes (0%–69%) No (70%–100%)	1–3	___ out of 6				
G-SRT.7	Explain and use the relationship between the sine and cosine of complementary angles.	___ out of 8		Yes (0%–69%) No (70%–100%)	4–7	___ out of 8				
G-SRT.8	Solve applied problems with right triangles using trigonometry and the Pythagorean theorem.	___ out of 15		Yes (0%–69%) No (70%–100%)	8–12	___ out of 12				

continued ↓

Figure 5.10: High school student tracker—Self-assessment and action plan.

Action Plan

Concept-Mastery Options I Will Attempt Prior to Taking a Recovery Quiz
(Check all that apply, and write dates and times: morning, lunchtime, advisory, or after school.)

Standards	Learning Target (I can...)	Standard Recovery Tutoring With Teacher	Standard Recovery Tutoring in the Mathematics Learning Center	Standard Recovery Tutoring With Another Teacher or Student	Standard Recovery by Independently Completing Review	Standard Recovery by Rereading Notes	Standard Recovery by Online Program	Date of Recovery Quiz	Score
G-SRT.6	Explain the definitions for trigonometric ratios for acute angles in right triangles using similarity.								
G-SRT.7	Explain and use the relationship between the sine and cosine of complementary angles.								
G-SRT.8	Solve applied problems with right triangles using trigonometry and the Pythagorean theorem.								

Student signature: _____ Date: _____

Parent or guardian signature: _____ Date: _____

Visit go.SolutionTree.com/MathematicsatWork for a free reproducible version of this figure.

Determine how to make the instructional minutes spent taking an exam be valuable learning minutes for students as they reflect, refine, and act (see page 2 in the introduction) on their feedback to inform future solution pathways and strategies.

TEACHER *Reflection*

What type of tracker can you and your colleagues create to prompt students to self-reflect on their learning after common mid-unit assessments?

_____
_____
_____
_____
_____
_____
_____

Self-Regulatory Feedback

Your collaborative team can also devise a system to help students generate their own self-regulatory feedback while taking a common formative check for understanding during a unit. Students not only solve the tasks on the assessment but also write the following symbols next to each item.

- **Plus symbol:** Next to items if they feel confident about having solved them well
- **Check mark:** Next to items if they feel partially confident about them and want to see, using feedback, the aspects of the solution they feel confident about and which parts might require additional learning
- **Question mark:** Next to items they were truly unsure about when taking the test but want feedback on to see if any part of their solution is on the right track

When students get their assessments back, they not only examine teacher feedback related to the mathematical tasks on the assessment but also check their solutions to grow their own self-regulatory feedback *during* the unit.

With the ease of scoring selected-response items using digital software, students should develop the same set of reflection skills when completing multiple-choice items. Anytime you ask students to solve multiple-choice mathematics tasks, ask the students to self-identify one of the following four strategies.

1. Just do it (read the question, solve it, and find the answer or answers)
2. Do it backward (look at the solutions and substitute them in to find the answer or answers)
3. Educated guess (eliminate one or more options and guess from there)
4. Pure guess (simply guess which option—or options—is the answer)

Next to each multiple-choice item, whether using clickers in class or taking a common formative assessment, students write *1*, *2*, *3*, or *4* to generate self-regulatory feedback about their choices for solving problems. Ideally, students begin to notice that strategies 1 and 2 should be chosen first to demonstrate learning and provide better feedback for continued learning.

These self-identification strategies work well when students are taking assessments on a computer or on paper if they are required to show their work and how they determined each answer. The work they show can then be used as part of the student reflection and action routine, once they receive feedback on the common digital assessment.

Use figure 5.11 (page 118) to determine how your teacher team will address feedback and student action during and after a unit from common formative assessments.

How will you help students identify strengths, weaknesses, and the essential learning standards they still need to meet for proficiency? What structures will you and they use to measure their progress? How do you expect students during the current unit or in the next unit to re-engage in learning as needed? How will your teacher team enrich or extend learning, as needed? These questions, when answered collectively as a team, and accompanied with quality feedback, guide the team to answer PLC critical questions 3 and 4 during the unit (DuFour et al., 2016): How will we respond when some students do not learn? and How will we extend the learning for students who are already proficient?

Directions: Use the following prompts to guide team discussion of your current teacher team agreements regarding feedback to students and student action to re-engage in learning.

1. How will we commit to calibrating our scoring and determine ways to give quality feedback to students?

2. What team structure or template will we commit to using with students to have them self-assess their learning and devise a learning plan, if needed, after the common end-of-unit assessment?

3. What team structure or template will we commit to using with students during a unit so they re-engage in learning by essential learning standard, as needed?

4. How will we develop students' agency as they take action on their feedback in this unit or the next?

5. What strategies will we use to support and model for students how to give each other meaningful feedback and learn from one another?

Figure 5.11: Teacher team discussion tool—Collaborative team feedback with student action.

*Visit **go.SolutionTree.com/MathematicsatWork** for a free reproducible version of this figure.*

TEAM RECOMMENDATION

Create a System for Student Action and Ownership of Learning Based on Formative Feedback During a Unit Using Common Mid-Unit Assessments

- Create a reflection tool for students to use throughout a unit as a way to monitor progress toward meeting essential learning standards before the unit ends.
- Require students to re-engage in learning if there are essential learning standards not yet understood.
- Determine how you can work together to minimize the number of students needing a Tier 2 intervention after the end-of-unit assessment by proactively working with students to learn the essential learning standards during the unit.

TEACHER Reflection

How can you utilize common mid-unit mathematics assessments as part of your mathematics team intervention program?

How can you utilize your common mid-unit mathematics assessments to extend student learning during a unit?

___

A strong formative feedback process requires students to use their common mathematics assessments for continued learning of essential learning standards. When this is done during the unit, gaps in student learning are more quickly closed, allowing students to be more prepared for learning in each subsequent unit.

Requiring students to re-engage in learning essential standards and learn from your feedback, as well as feedback from peers, is generated and supported through the opportunities for intervention that your teacher team and school provide. In the next chapter, you will examine the criteria of highly effective mathematics intervention programs.

Visit **go.SolutionTree.com/MathematicsatWork** for free reproducible versions of tools and protocols that appear in this book, as well as additional online only materials.

CHAPTER 6

Team Response to Student Learning Using Tier 2 Mathematics Intervention Criteria

As long as interventions are viewed as an appendage to a school's traditional instructional program, instead of an integral part of a school's collaborative efforts to ensure all students succeed, interventions will continue to be ineffective.

—Richard DuFour

In chapter 4, you learned about team data-analysis routines that contribute to determining your next instructional steps for students and for an analysis of your most effective instructional practices. In chapter 5, your team explored strategies to engage students in their learning after common mid-unit and end-of-unit assessments using reflection and action tools. Both become important routines in using your common assessments formatively.

The next step is to decide how you, your teacher team, and your students will use assessment results and evidence from student work to intervene and continue learning the essential standards for each mathematics unit.

In other words, *now what*?

The last criterion in the formative assessment process rubric in figure 1.2 (page 12) is to develop an effective Tier 2 intervention program.

Not all students learn the same way or at the same rate, as evidenced by the student work on your common end-of-unit mathematics assessments. When students have not yet learned essential learning standards, the collaborative teacher team response in a PLC at Work culture is *not* to stop and reteach all students. Such an action would be at the expense of students' learning the remaining essential standards that are part of the guaranteed and viable curriculum promised for your grade level or mathematics course.

Instead, the PLC at Work response when students are not learning is to work as a team, to design a quality mathematics intervention program that provides continued support toward learning essential grade-level or course-based standards, while simultaneously teaching the essential learning standards of the next unit. Such a mandate requires a team effort to ensure every student is learning from an intentional mathematics intervention. Quality systematic interventions are not pull-out programs or initial differentiated instruction, but rather a value-added model of continued instruction and student interaction.

There are generally three tiers of mathematics intervention, which are included in any response to intervention (RTI) system of supports (your school, district, state, or province may use RTI or multitiered system of supports [MTSS] as an intervention framework). RTI provides interventions for students during core instruction (Tier 1), in small identified groups using additional time during the school day for intervention (Tier 2), and as individuals for even more intentional acceleration of learning to grade-level standards (Tier 3). (For more information on the three tiers of RTI, visit AllThingsPLC [www.allthingsplc.info/blog/view/335/connecting-plcs-and-rti] to read the blog of RTI expert Mike Mattos [2016].)

Tier 2 mathematics interventions target students in your grade level or course who are *not yet* meeting proficiency for the essential learning standards in a given mathematics unit and who are in need of additional time and support outside of the daily lesson structure.

The instruction provided during any Tier 2 intervention time should be targeted and specific to the learning targets and essential learning standards from previous and current mathematics units.

It is possible that as you and your colleagues use high-quality instructional practices in Tier 1, as described in *Mathematics Instruction and Tasks in a PLC at Work, Second Edition*, it will help limit the number of your students who require the teacher team–developed, Tier 2–type mathematics interventions described in this chapter.

For the purposes of this book related to your teacher team's common assessments and interventions, the focus is on Tier 2–type interventions. The intent of your mathematics interventions should be to provide students with additional time and support needed to learn your grade-level or course-based essential learning standards. NCTM (2020) shares that students should be provided the additional time and support using assessments "connected to students' coursework" (p. 96)—in essence, your team's common unit-by-unit mathematics assessments.

Figure 6.1 shows the various times when your teacher team may respond to student learning on common assessments during the unit after a common mid-unit assessment and after the common end-of-unit assessment. While figure 6.1 shows students re-engaging in learning the essential standard or standards from one unit to the next, at times Tier 2 may also include a team response to a common mid-unit assessment.

Developing a teacher team Tier 2 response to student learning can be difficult when colleagues are not yet comfortable with data-analysis protocols to determine the contents of the Tier 2 intervention.

In Mona Toncheff's story, the teachers at Field Elementary asked students to self-identify their areas of strength and weakness as part of the WIN process. Additionally, the teacher team established a viable and collaborative Tier 2 response that allowed students to take action on their initial assessment performance and receive help and support from a wide range of faculty and staff.

The questions in figure 6.2 (page 124) serve to help your team members understand one another's perspectives related to your systematic interventions as a teacher team. In other words, how does your teacher team respond when common assessment data reveal that some students have learned the essential learning

> **TEACHER *Reflection***
>
> When a common end-of-unit mathematics assessment shows that students have not learned essential learning standards, what is *your* response?
>
> What is your *teacher team's* response?
>
> What do you require all students in the grade level or course to do?
>
> _____
> _____
> _____
> _____
> _____
> _____
> _____

standards and others have not? Your professional response as individual teachers and as a team will reveal your current beliefs about the need for interventions and the plan for making them effective.

Notice question 5 addresses a needed team discussion related to student grades. When grades provide feedback on student learning and students demonstrate learning later, teacher teams agree on how to ensure the grade affects student learning of the standards. In other words, if students can demonstrate learning after retaking an initial high-quality assessment following additional learning, how is that grade used as part of the student's overall grade since it best reflects student learning of the essential standards on that assessment? (For more information, see *Mathematics Homework and Grading in a PLC at Work*.)

Use the self-evaluation tool in figure 6.3 (page 124) to rate the quality of your teacher team response to intervention and learning evidenced on a recent common end-of-unit assessment. How does your current mathematics intervention program score? You should expect to develop an intervention program that scores 4s in all five criteria.

Each of the five criteria is needed for continued student learning when a student has not yet learned an essential learning standard, as evidenced on the common end-of-unit mathematics assessment.

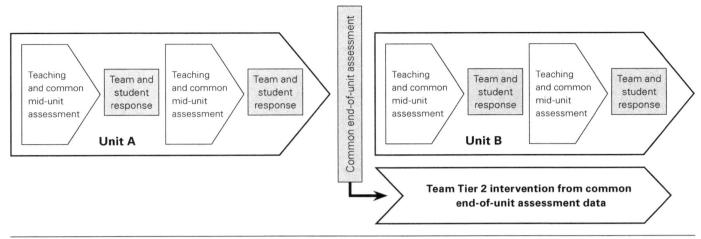

Figure 6.1: Common assessment formative process for teachers and students.

Personal Story 66 MONA TONCHEFF

When teams analyze their common assessments, the biggest hurdle to overcome is to plan *now what?* When I was working with K–5 mathematics teams at Field Elementary in Louisville, Kentucky, the faculty and staff worked diligently to collectively answer the *now what?* question.

The collaborative teams met every Friday to review the common assessment data from the week and identified students in each grade level who were meeting, exceeding, or struggling with the essential standards. The teachers on each team decided they needed to create a plan for re-engagement or enrichment for students during their designated WIN (**W**hat **I** **N**eed) time.

When the grade-level teacher team started this process, the teachers were reluctant to share their students' information during WIN time. The first year, some teachers would keep their students in their classrooms and try to differentiate mathematics instruction using small student groups. However, demonstrations of student learning were not improving. Students with learning differentials during the second week were still struggling with the same concepts in the twelfth week.

By midyear, the teacher teams were more willing to discuss instructional practices that were shown as most effective to have students learn the agreed-on essential learning standards. They started rotating students to different teachers during WIN time to better target specific instructional practices or student strategies that would move students forward by standard. Student progress was continually monitored, and student movement was fluid and flexible during WIN time. Students also started identifying what they had learned or not learned yet after common assessments.

By the second year, the entire staff was also part of the solution. Support staff were assigned to designated groups or teams during WIN time. They also were part of the team and participated during team time to understand how to address the learning needs and support the learning outcomes for each student. The common assessment evidence drove their WIN time schedule, and teams were changing the status quo of student learning to address students' diverse learning needs. See AllThingsPLC's (n.d.) "Evidence of Effectiveness: Field Elementary" (www.allthingsplc.info/evidence/details/id,998) to read more about the school's success.

Directions: Use the following prompts to guide team discussion of your current collective team intervention practices.

Plan for team intervention:

1. How does your teacher team determine the targeted content and skills to address through Tier 2 interventions and the students who need each?

2. Who provides the Tier 2 interventions for students in your class or across your teacher team? When? How do you and your colleagues share the responsibility of interventions for students across your grade level or course?

3. How do students move in and out of interventions? How often?

Effectiveness of interventions:

4. How does your teacher team know if your interventions are effective?

5. How are students' grades impacted when they show learning after effective interventions?

Figure 6.2: Teacher team discussion tool—Intervention practices.

Visit go.SolutionTree.com/MathematicsatWork for a free reproducible version of this figure.

The challenge is to create an effective system that allows students to engage in the intervention, while simultaneously practicing the essential learning standards for the next unit. Your teacher team must attend to your schedule's structure in order to determine when to provide Tier 2 supports as well as address the targeted content needed to accelerate student learning to grade level and beyond.

As described in figure 6.3, there are five criteria your teacher team must consider when building your mathematics intervention program as part of your team's formative assessment process (surfing).

1. Systematic and required
2. Targeted by essential learning standard
3. Fluid and flexible
4. Just in time
5. Proven to show evidence of student learning

The following sections explain each of the five criteria in more detail.

TEACHER *Reflection*

Which of the five criteria for a high-quality Tier 2 mathematics intervention program are currently part of your collaborative team practice?

What do you need to do to strengthen your mathematics intervention program?

_____
_____
_____
_____
_____

Mathematics Program Tier 2 Criteria	Description of Level 1	Requirements of the Indicator Are Not Present	Limited Requirements of the Indicator Are Present	Substantially Meets the Requirements of the Indicator	Fully Achieves the Requirements of the Indicator	Description of Level 4
1. Systematic and required	Teachers provide optional opportunities, often before or after school, for students to engage in individual or small-group help.	1	2	3	4	Teachers on the team require students to learn if they are not yet proficient with essential learning standards and provide systematic structures during the school day to ensure student learning. The team commonly plans for continued learning in the next mathematics unit, as needed.
2. Targeted by essential learning standard	Teachers work with individual students or try to work with many students needing different interventions at the same time. Focus may be on a grade or missed assignments.	1	2	3	4	Teachers on the team analyze data for all students and collectively respond to the data to identify specific students needing targeted intervention by each essential learning standard.
3. Fluid and flexible	Teachers place students in an intervention due to results from a diagnostic assessment, and students stay in their group for a long period of time with no movement when they learn.	1	2	3	4	Teachers on the team regularly analyze student data from common unit mathematics assessments and move students in and out of the required additional time and support in learning as needed.
4. Just in time	Students only receive help before the end of a grading period or when they request help.	1	2	3	4	Students get real-time feedback, and the teachers on the team plan interventions for the essential learning standards within a mathematics unit or at the start of the next unit using current common assessment results.
5. Proven to show evidence of student learning	Teachers provide intervention to students and hope students learn. Students may be receiving intervention support from a person not highly qualified in mathematics. There is little evidence that the intervention is helping students learn the essential standards for the grade level or course. Interventions may be more focused on steps and algorithms.	1	2	3	4	Teachers on the team monitor student progress within the intervention or within class to see if the intervention is effective and results in student demonstrations of proficiency for each essential standard. Each adult implementing the team-designed mathematics intervention is the best-qualified person for the role. Team members monitor instructional strategies that impact learning and re-engagement and include conceptual understanding in any intervention.

Figure 6.3: Teacher team rubric—Tier 2 mathematics intervention program.

Systematic and Required

When your teacher team plans for additional time and support for students who still need to learn essential learning standards, time is often an issue. And until you address issues of time, you will struggle to have a strong Tier 2 program that is both systematic and required. Your teacher team can solve some of these time issues, but systematic interventions require leadership and a designated time during the school day in the master schedule.

To be systematic, your mathematics intervention program needs a consistent structure within the school day. To be required means that the students your teacher team identifies as needing intervention cannot opt out of the additional learning. Figure 6.4 shares some options to account for Tier 1 instructional time and Tier 2 interventions and extensions weekly.

In figure 6.4, option A is ideal with time each day for uninterrupted core instruction and additional time built into the day for Tier 2 interventions and extensions. The Tier 2 additional time might be at the same time across the school, providing many opportunities to regroup students by targeted skill and include all educators on campus. The Tier 2 additional time may also be at different times of the day due to a grade-level team's schedule or the need to use different periods at the high school level (for example, pulling students in mathematics for Tier 2 intervention during an assigned study hall). In tandem with option A, or when option A does not yet exist, team members can plan a flex day when planning their unit and preplan for the need to re-engage students in learning (shown in option B). In option C, the team designs a collective Tier 2 response that is employed during small-group instruction time. In option D, teachers teach their Tier 1 lesson in a shorter period of time to create two thirty-minute blocks of time each week for interventions. Students can be shared between teachers in any of these intervention models, or, if it is impossible to share due to the timing of class periods or the space between classrooms, teachers on your collaborative team can each design an intervention activity and share their activities with one another to implement in stations.

You should recognize that some students are not learning the essential standards because of a lack of effort, and they choose failure. As Timothy Kanold indicates in his story on the next page, those students are choosing not to mow the lawn. Allowing such irresponsibility does not teach your students to become more responsible.

Possible Schedule	Monday	Tuesday	Wednesday	Thursday	Friday
A	Tier 1 Instruction	Tier 1 Instruction	Tier 1 Instruction	Tier 1 Instruction	Tier 1 Instruction
	Tier 2 Intervention	Tier 2 Intervention	Tier 2 Intervention	Tier 2 Intervention	Tier 2 Intervention

Or

Possible Schedule	Monday	Tuesday	Wednesday	Thursday	Friday
B	Tier 1 Instruction	Common Assessment	Flex Day	Tier 1 Instruction	Tier 1 Instruction

Or

Possible Schedule	Monday	Tuesday	Wednesday	Thursday	Friday
C	Tier 1 Instruction	Tier 1 Instruction	Tier 1 Instruction	Tier 1 Instruction	Tier 1 Instruction
		Small Group	Small Group	Small Group	

Or

Possible Schedule	Monday	Tuesday	Wednesday	Thursday	Friday
D	Tier 1 Instruction	Tier 1 Instruction	Tier 1 Instruction	Tier 1 Instruction	Tier 1 Instruction
		Thirty-Minute Intervention		Thirty-Minute Intervention	

Figure 6.4: Tier 1 and Tier 2 schedule options.

Personal Story TIMOTHY KANOLD

When Rick DuFour and I first began the process of required interventions at Stevenson High School in Lincolnshire, Illinois, we would often use the following lawn-mowing story to make our point.

If I asked my son to mow the lawn by Friday evening, and I came home and he had not yet mowed the lawn or had mowed it poorly, my response would not be, "Well, since you chose not to mow the lawn, or at best you mowed it poorly, you will never have to mow the lawn again." Instead, my response would be, "You will mow the lawn, and you will mow it to my satisfaction, or you will not be allowed to see your friends this weekend!"

It eventually became part of the folklore at Stevenson. The analogy of course is that this is what we tend to do with students and our assessments. When there are standards on an assessment for which students show they are not yet proficient, our response must never be, "Because you did not learn them by this Friday, you will never have to learn these standards."

A required student action to learn is vital for a strong mathematics teacher team intervention response. Students do not get the choice of failure, nor do they get to avoid re-engaging in learning the essential learning standards.

Think about the current intervention opportunities for students at your school. Now ask yourself, How many of the interventions does your school require? Do students who need the intervention attend or participate in the intervention?

How do your teacher team and school embrace the PLC at Work culture to make it easier to pass than to fail? A required response means the intervention happens during the time students must be on campus and your collaborative team ensures students attend and participate in the learning opportunity. As you work together to determine appropriate interventions, your teacher team should also ensure each academic intervention is not happenstance.

Your collaborative team's response to student learning should address the following questions.

- Is the intervention needed for a few students or several students, or is there a need to address learning for all students?
- Which essential learning standards do students still need to learn? Which specific mathematical concepts, skills, or applications do they still need to learn?

TEACHER Reflection

Write *required* or *not required* following each sample intervention routine that may apply for your grade level or course.

- Mandatory study hall _____
- Homework help sessions _____
- Peer-tutor program _____
- Parent- or community-student support teams _____
- Lunchtime learning or other program that shows student learning is not optional _____
- Walk-to-mathematics during the school day (elementary) _____
- Additional mathematics-support class (secondary) _____
- Additional time built into the master schedule for interventions and extensions _____
- Use of a mathematics learning center on campus for student assignment to receive help from a mathematics teacher or peer tutor by unit and standard _____
- Assembly bell schedule one or two days per week; assembly time is used for Tier 2 intervention _____

- Is the intervention tailored to meet the needs of a specific student population, if needed?
- What accommodations or modifications are you implementing to differentiate learning for students when it is required?
- What responsive teaching do you need for all students to be successful?
- What manipulatives or technologies should you use to develop greater conceptual understanding and procedural fluency for each standard?

As you work to answer these questions, use the template in figure 6.5 to help you organize your thinking and create a collective team plan of mathematics intervention your students can access.

Depending on the number of students needing intervention with an essential learning standard, your teacher team might also determine a plan to weave a required intervention into the instructional time it allocates to the next unit. Addressing intervention in the next unit, or before the next unit begins, occurs when many students across the team have not yet learned an essential standard. Using instructional time in the next unit is also a possible response if your school does not yet have a time built into the master schedule for intervention in addition to core instruction (see also figure 6.4, page 126). The intervention might be:

- A lesson or series of lessons the team agrees to implement
- A flex day the team uses to share students, with each team member addressing a particular skill that students need to learn
- A concept intentionally knitted into specific and agreed-on lessons for the next unit (for example, continue working on solving word problems using team agreed-on strategies)
- Beginning-of-class routines focused for two weeks on engaging students in learning the essential learning standard

For additional ideas related to your daily schedule and how to build in time for intervention, visit www.allthingsplc.info/evidence and look at the schedules from the featured Model PLC schools.

Overall, you can strengthen a Tier 2 mathematics program when it is a normal part of the school day or week and when you require students to re-engage in learning across the grade-level or course-based team.

> **TEACHER Reflection**
>
> How will you and your colleagues work with your school administrators to design a mathematics intervention program that is systematic and required?

Targeted by Essential Learning Standard

After your teacher team analyzes the data from a common unit assessment, you will determine specifically which students are proficient with each essential learning standard and which are not. (See chapter 4, page 81, for common assessment data-analysis protocols.) You also need to identify trends in student reasoning that must be corrected.

Your teacher team can then create an intervention plan to determine which students need to re-engage in learning each essential learning standard and how to most effectively target that learning. Ideally, through self-reflection, students might be able to tell you which essential learning standard they most need to work on during the intervention (see chapter 5, page 97).

Mathematics interventions vary in productivity or consistency for students when teachers make individual decisions about how to respond to learning. The learning is not equitable for all students. The interventions must reflect your teacher team response to student learning so each student in the grade level or course receives the same access to learning the team-developed expected outcomes. This is an equity issue for you and your colleagues.

Anytime your systematic mathematics intervention time occurs and there are some students who have learned the essential standards in the unit, your teacher

Essential Standard:

What Is the Mathematics Intervention?	Which Students Need Intervention?	How Often Will Students Get Support?	How Will Students Engage in Intervention?

Figure 6.5: Teacher team discussion tool—Mathematics intervention planning tool.

Visit go.SolutionTree.com/MathematicsatWork for a free reproducible version of this figure.

team will also have to plan for targeted extensions. Determine a targeted extension that does not require students to peer-tutor but rather challenges each student to continue on their learning journey.

You and your colleagues should create quality mathematics interventions focused on a targeted essential learning standard or skill. In a PLC at Work culture, your team works together to determine the most effective intervention plans to help each student respond to the feedback they receive from your common unit mathematics assessments.

TEACHER *Reflection*

How will your mathematics intervention program group students by targeted skill so students can re-engage in learning a specific part of an essential learning standard equitably across your teacher team?

_____
_____
_____
_____
_____
_____
_____

In addition to being systematic and required and targeted to specific essential learning standards, an effective mathematics intervention program must also meet the learning needs of students in a fluid and flexible way.

Fluid and Flexible

A third criterion for a quality mathematics intervention program is that students fluidly and flexibly move into and out of needing additional learning supports once your teacher team establishes the mathematics interventions you plan to provide students at the end of every unit. There are two critically important questions you must ask and answer based on student assessment results.

1. How do we provide access for students to the mathematics intervention program?

2. How do we help students exit the mathematics intervention program once they exhibit proficiency on the targeted deficient mathematics standards?

Your expectations for mathematics intervention should not permanently place your students in a systematic and required intervention. Once students have demonstrated mastery of the expected mathematics learning standards, you should remove the intervention. Moving forward, the daily mathematics lesson time or class should be sufficient.

For example, if you are working with elementary students in a walk-to-mathematics program or high school students in a mathematics tutoring room or study hall or during additional time built into your master schedule, you may be able to target students after a unit assessment for a two- to three-week period of time. Meanwhile, your team uses Tier 1 core instruction to continue the learning of mathematics with the next unit. Once your students have learned the essential learning standards, you should remove them from additional support either at the end of the designated time period or before.

If, however, part of your Tier 2 mathematics intervention program in middle school or high school includes an additional class period of support, moving students into and out of the support is not as fluid or flexible due to other constraints on students' schedules. This is something you and your colleagues will need to address, most likely with the support of your principal, as students need to know the mathematics intervention is not forever and can be removed from their schedule once they reach proficiency.

Moving students into and out of your intervention program is essential if students are going to have additional time and support to learn essential learning standards when that learning is most critical. Consider, too, Sarah's story on the next page and the importance of not labeling your students in need of intervention.

Yet, all this effort will be for naught unless you can re-engage your students in targeted learning soon after any common unit assessment. It is important to move students out of intervention when they have learned, and also into intervention when they have not learned yet.

Your students should know that the purpose of the intervention is to help them demonstrate learning the mathematics standards—just in time.

Personal Story SARAH SCHUHL

A second-grade student stood to explain a subtraction problem that required regrouping to the class. Later, the student's teacher asked if I had noticed the student's thorough explanation. She marveled at his ability to subtract and exclaimed, "Can you believe it? He's one of my Tier 2 students!"

I simply replied, "Well, he's not Tier 2 related to subtraction."

In fact, no one is a Tier 2 or Tier 3 student. These are not labels of learning but rather structures to make sure students are getting the additional supports they need to learn.

TEACHER Reflection

How often will you move students into and out of your Tier 2 mathematics interventions and supports?

How will you structure your intervention program to make it fluid and flexible?

How much choice do students have for the nature of the intervention and support they receive?

_____
_____
_____
_____
_____

Just in Time

As previously described, a Tier 2 mathematics intervention program should be systematic and required, targeted by essential learning standard, and fluid and flexible. Additionally, to be effective, the program must provide students with just-in-time feedback in order to grow their learning and give students access to learning grade-level or course-based content. This means the intervention focuses on essential learning standards students are currently learning or have just learned in the previous mathematics unit.

When a Tier 2 intervention program only uses district benchmark assessments or universal screeners to determine who should be in the intervention program, several problems for student learning may occur. Universal screeners and benchmark assessments seldom provide feedback specific to targeted skills or standards that students need to learn (not targeted by essential learning standard). Additionally, educators don't give universal screeners and benchmark assessments frequently enough to allow for a fluid and flexible program. Nor are universal screeners and benchmark assessments able to give just-in-time feedback to students.

While universal screeners and district benchmark assessments do provide additional data and can be informative, they alone do not give your teacher team the most effective data for a quality Tier 2 program, nor are those mathematics assessments generally used for formative student learning purposes. Universal screeners are used most effectively to determine the students in need of Tier 3 supports. It is only through your *team-created common assessments* that you give students just-in-time feedback and allow your Tier 2 program to re-engage students in learning necessary for success in that moment of time.

When Tier 2 interventions are *not* tied to current or recent essential learning standards, students sometimes miss the purpose of re-engagement in learning because it lacks connections. Using a computer program alone for interventions when not assigning targeted essential mathematics standards for students to complete in the program does not provide feedback related to their current learning. Too often, this type of intervention content causes student confusion and teacher frustration because students do not necessarily grow in their learning despite the time spent in intervention.

Similarly, addressing standards from several grade levels before is not advantageous to students' learning the current essential learning standards and would fall into a Tier 3 need for students.

A just-in-time, systematic Tier 2 mathematics intervention program does not ask students to get help weeks after learning; rather, such a program has students get help right away. The National Council of Supervisors of Mathematics (NCSM, 2020) shares that student Tier 2 interventions should be at grade or course level related to essential learning standards or to the immediate prior knowledge standards leading to the essential learning standards.

> **TEACHER *Reflection***
>
> How does your current mathematics intervention provide just-in-time learning of the essential learning standards students have not learned yet but have addressed in the current or previous unit?
>
> _____
> _____
> _____
> _____
> _____
> _____
> _____
> _____

Finally, your teacher team will only know if a Tier 2 mathematics intervention program is successful if there is evidence of student learning.

Proven to Show Evidence of Student Learning

Once your teacher team determines a targeted intervention plan, you must also determine if the intervention was effective. Did the intervention improve student learning in the targeted essential learning standard or skill? What evidence of success will help your team to know?

When determining effectiveness, further demonstrations of student learning are most important. Your colleagues will need to ask how the instructional strategies used and the teacher who employed the intervention impacted student learning in current in-class work.

NCTM (2014) advocates that quality Tier 2 mathematics intervention learning opportunities require teachers who are most skilled in teaching mathematics. The most skilled teachers for Tier 2 interventions are usually the professional educators teaching that grade level or course.

In a PLC at Work culture, your teacher team does not think of these Tier 2 interventions as places or people to send students to be fixed. Rather, Tier 2 intervention for students is a program and process owned by the teacher team for greater effectiveness. The focus for effective Tier 2 interventions is first ensuring students understand the concept of the essential learning standard and its connections to prior learning using a variety of strategies and tools, versus repeating the same "steps" or strategies to solve tasks with algorithms.

Additionally, mathematics support and interventions in Tier 2 should look significantly different from the students' initial learning experience (NCSM, 2015). For example, students in first grade who have not learned addition yet, even with the supports of linking cubes or ten frames, may need an alternative strategy to build conceptual understanding. Instead, students may need to play board games and spin a spinner or roll a die (and subitize) before moving a piece the designated number of spaces. Students discuss subtraction when determining how many spaces are needed to win and addition when looking at the total number of spaces traveled.

In later grades, a Tier 2 intervention focusing only on reteaching an algorithm does not often produce more learning. Developing a conceptual understanding for the algorithm and connecting it to previous learning better grows student understanding and application during the intervention time. For example, students needing intervention for an essential standard on graphing linear functions may need to understand the concept by graphing several linear functions using technology and connecting the linear function graphs to proportional relationships learned in seventh and eighth grade.

Students still in need of learning basic facts should experience efficient, effective, and flexible strategies for mental use and application, as memorizing an answer is not a successful pathway for them (see *Mathematics Instruction and Tasks in a PLC at Work, Second Edition*).

When looking at your common assessment results, your teacher team should determine best instructional practices that result in student learning and use that information to strengthen both Tier 1 instruction and Tier 2 interventions. During this time, your teacher team may want to determine best uses for manipulatives and other technologies or digital resources. How do you give students access to these resources, and how do students use them for explorations to deepen conceptual understanding or to visualize or make sense of real-life applications and modeling? How are students "seeing" the mathematics instead of simply manipulating numbers and symbols?

As your teacher team designs Tier 2 intervention experiences tied to essential learning standards, consider who is the best person to lead each group of students and how students should be re-engaged in learning the content. Then, consider how you will know whether or not students actually increased their learning because they engaged in the intervention. Consider also how students will know whether or not they have learned the targeted skill in the intervention and reflect on their progress toward learning the essential standard.

You will want to use short, team-created common assessments during the mathematics intervention to see if students have learned the essential mathematics standard for which they are receiving additional support. This intervention assessment should provide opportunities for students to retake all or part of a common mid-unit or end-of-unit assessment. Knowing whether or not an instructional strategy improved learning and whether or not students have grown in their learning due to the intervention you provided is essential to an effective Tier 2 mathematics intervention program.

Mathematics intervention is about creating equity and access to meaningful mathematics learning for your students. Tier 2 intervention does not mean placing all students with learning difficulties in the same classroom and pulling them out of core instruction. Tier 2 *is not* about leveling students (low, middle, high) at the elementary or secondary level and giving them the same amount of time to learn as a lower-ability group for first-best instruction. This type of homogeneous grouping is harmful to your students.

NCTM (2014) explains, "The practice of isolating low-achieving students in low-level or slower paced mathematics groups should be eliminated" (p. 64). The Tier 2 intervention should provide *additional* time and support beyond core instruction (daily lessons) to learn the essential standards so every student learns the essential learning standards at grade-level understanding or above. It is during Tier 2 interventions students are put into small groups addressing a targeted and specific learning target in an essential learning standard.

Kanold and Larson (2012) advocate that effective intervention programs should:

- Be mandatory, not optional (i.e., scheduled during the school day whenever possible);
- Be based on constant monitoring of students' progress, as determined from the results of formative and summative assessment, ensuring that students get support as quickly as possible;
- Attend to conceptual understanding as well as procedural fluency; and
- Allow for flexible movement in and out of the intervention as students need it (p. 66)

Regardless of the structures you have in place to provide additional time and support, your collaborative team members will need to navigate the complexities of the school day and collectively respond to the evidence of student learning by requiring students to participate

---

TEACHER *Reflection*

Who provides the Tier 2 mathematics intervention your students need?

How do you decide if the intervention is effective for closing gaps in student learning of the essential learning standards? What data do you examine to decide this?

in coordinated intervention efforts. The interventions you provide should not be a student choice. As soon as a mathematics unit ends, students who are not yet able to demonstrate learning of the standards should be immediately required to engage in mathematics intervention and support. Thus, you should design your feedback for assignments, assessments, and classwork to allow students to create an action plan.

Imagine preK–12 students who are struggling in mathematics. Are they allowed to give up, stop trying, or take an "easy" zero? Or do they know your team has their back? Do the students know there are guaranteed supports in place that require them to accelerate their learning and succeed because their teachers believe in them and know they can learn at high levels?

TEAM RECOMMENDATION

Create a Tier 2 Mathematics Intervention Program

Use common assessments to create a team response to student learning that is:

- Systematic and required
- Targeted by essential learning standard
- Fluid and flexible
- Just in time
- Proven to show evidence of student learning

Determine how often your teacher team will meet to plan for an effective Tier 2 intervention response and how you will hold one another accountable to your team intervention plan.

A quality, team-created Tier 2 mathematics intervention program erases the inequities team members create when they design isolated interventions and fail to be transparent with one another. You should make your end-of-unit intervention response a priority for the work of your teacher team.

Unsurprisingly, the formative assessment process becomes stronger if your students do not wait until after the common end-of-unit assessment to self-reflect. When students reflect after common mid-unit assessments, the continual reflection allows for additional learning as needed during a unit so students can more confidently demonstrate evidence of learning on the end-of-unit assessment.

In chapter 3 (page 47), you examined calibrating common assessment scoring and providing students with consistent FAST (fair, accurate, specific, and timely) feedback as part of a strong formative assessment process. Your teacher team uses the information revealed in the student work as part of their data-analysis and action routines to identify targeted and specific interventions needed by groups of students across the team (chapter 4, page 81). Students then use your FAST feedback to self-reflect and set goals for continued learning (chapter 5, page 97). Your team's robust Tier 2 mathematics intervention program provides the resources and support students need for that continued learning. These teacher team and student action routines create a meaningful and productive use of your team-created common assessments.

Together, as a collaborative team, you have created a formative process for student learning. The reward for your effort will be significant improvement in your students' learning of mathematics and their confidence in that learning process along the way.

Visit **go.SolutionTree.com/MathematicsatWork** for free reproducible versions of tools and protocols that appear in this book, as well as additional online only materials.

Summary

How can you and your colleagues ensure every student accelerates their learning to grade or course level expectations in mathematics? Are you willing to extend the learning for each student in your grade level or course through a strong formative assessment process? In such a process, you provide each student with more equitable learning expectations and the necessary scaffolding to meet those expectations.

Using the common assessment formative process rubric in figure 1.2 (page 12), your teacher team addresses the six criteria needed for effective and actionable feedback and learning for both your teacher team and the students. Your work as a teacher team starts with clear essential learning standards for the unit and high-quality common assessments with agreed-on scoring agreements.

Next, recall the questions that have repeatedly been asked throughout this book: How does your teacher team analyze the common assessment results and collectively respond? What happens in the classrooms across your collaborative team when you return a common mathematics assessment to your students? What do you expect students to do with the feedback you provide? An assessment instrument is meaningful and useful only when it becomes part of a formative assessment process of learning you and your colleagues establish.

The following four teacher team actions, also part of the common assessment formative process rubric in figure 1.2, are important in developing a quality formative assessment process.

1. Calibrate your scoring of common assessments by essential learning standard so students get consistent FAST feedback from which they can learn.

2. Analyze student learning data from common assessments to identify effective instructional practices and create a targeted team plan for continued student learning.

3. Determine how students will self-reflect and set goals for learning based on feedback on common end-of-unit and mid-unit assessments.

4. Create a quality Tier 2 mathematics intervention program that requires students to re-engage in learning or extend just-in-time learning based on the essential learning standards.

Taken together, your intentional common assessment design, calibration routines with feedback, data analysis of student work, and structures for student reflection and re-engagement along with a strong intervention program create the formative assessment process linked to gains in student learning (Gallimore et al., 2009).

Your teacher team has addressed the two critical team actions related to mathematics assessment and intervention.

- **Team action 1:** Develop high-quality common assessments for the agreed-on essential learning standards.
- **Team action 2:** Analyze and use common assessments for formative student learning and intervention.

Thus, students are able to *reflect*, *refine*, and *act* over and over and then over again. In short, they actively engage in the practice of learning mathematics.

Appendix

Cognitive-Demand-Level Task Analysis Guide

This appendix provides criteria you can use to evaluate the cognitive-demand level of the tasks you choose each day. As you develop your unit plans, use this tool (table A.1) to ensure a balance of lower- and higher-level tasks for students as they learn the standards.

Table A.1: Cognitive-Demand Levels of Mathematical Tasks

Lower-Level Cognitive Demand	Higher-Level Cognitive Demand
Memorization Tasks • These tasks involve reproducing previously learned facts, rules, formulae, or definitions to memory. • They cannot be solved using procedures because a procedure does not exist or because the time frame in which the task is being completed is too short to use the procedure. • They are not ambiguous; such tasks involve exact reproduction of previously seen material and what is to be reproduced is clearly and directly stated. • They have no connection to the concepts or meaning that underlie the facts, rules, formulae, or definitions being learned or reproduced.	**Procedures With Connections Tasks** • These procedures focus students' attention on the use of procedures for the purpose of developing deeper levels of understanding of mathematical concepts and ideas. • They suggest pathways to follow (explicitly or implicitly) that are broad general procedures that have close connections to underlying conceptual ideas as opposed to narrow algorithms that are opaque with respect to underlying concepts. • They usually are represented in multiple ways (for example, visual diagrams, manipulatives, symbols, or problem situations). They require some degree of cognitive effort. Although general procedures may be followed, they cannot be followed mindlessly. Students need to engage with the conceptual ideas that underlie the procedures in order to successfully complete the task and develop understanding.
Procedures Without Connections Tasks • These procedures are algorithmic. Use of the procedure is either specifically called for, or its use is evident based on prior instruction, experience, or placement of the task. • They require limited cognitive demand for successful completion. There is little ambiguity about what needs to be done and how to do it. • They have no connection to the concepts or meaning that underlie the procedure being used. • They are focused on producing correct answers rather than developing mathematical understanding. • They require no explanations or have explanations that focus solely on describing the procedure used.	**Doing Mathematics Tasks** • Doing mathematics tasks requires complex and nonalgorithmic thinking (for example, the task, instructions, or examples do not explicitly suggest a predictable, well-rehearsed approach or pathway). • It requires students to explore and understand the nature of mathematical concepts, processes, or relationships. • It demands self-monitoring or self-regulation of one's own cognitive processes. • It requires students to access relevant knowledge and experiences and make appropriate use of them in working through the task. • It requires students to analyze the task and actively examine task constraints that may limit possible solution strategies and solutions. • It requires considerable cognitive effort and may involve some level of anxiety for the student due to the unpredictable nature of the required solution process.

Source: © 1998 by Smith & Stein. Used with permission.

Mathematics Assessment and Intervention in a PLC at Work Protocols and Tools

Team Action 1:
Develop high-quality common assessments for the agreed-on essential learning standards.

Team Action 2:
Analyze and use common assessments for formative student learning and intervention.

Actions Teacher teams must . . .	Questions the Protocol or Tool Will Address	Protocol or Tool
Chapter 1: The Mathematics at Work™ Common Assessment Process		
Answer questions about their current use of common assessments in terms of feedback to students, student actions, and teacher actions.	How quickly do students receive their assessment results? How do students learn from their assessment results? How do teachers respond to student learning?	Figure 1.1: Teacher team discussion tool—Formative use of common assessments (page 10)
Analyze and evaluate the team formative actions related to common assessments.	Using the rubric, how does your team score for team formative processes using common assessments? What is a strength? What needs to be addressed?	Figure 1.2: Teacher team rubric—Common assessment formative process (page 12)
Discuss current design and use of common assessments.	How is the team planning for effective common assessments? What are the teachers on the team and students doing with the data from common assessments?	Figure 1.3: Teacher team discussion tool—Common assessment self-reflection protocol (page 14)
Chapter 2: Quality Common Mathematics Assessments		
Analyze and evaluate the quality of common assessments.	Using the rubric, how does your team score for the quality of your common assessments? What is a strength in team assessment design? What needs to be revised on common assessments?	Figure 2.1: Teacher team rubric—Assessment instrument quality evaluation (page 18) Figure 2.2: Teacher teach discussion tool—High-quality assessment evaluation (pages 19–20)
Determine essential learning standards to be used in a unit for assessment and reflection.	How does the team unwrap standards to determine what students should know and be able to do? What will be the student-friendly targets used for essential learning standards (ideally, five or fewer per unit)?	Figure 2.3: Sample grade 4 fractions unit—Essential learning standards (pages 24–25)
Identify lower- and higher-level-cognitive-demand tasks for standards.	What is the difference between a lower-level task and a higher-level task? What are examples of lower- and higher-level tasks to the same standard?	Figure 2.4: Examples of lower- and higher-level-cognitive-demand tasks (pages 27–28)

Practice scoring agreements for assessment tasks. *Note: You can use team assessments for this exercise with figure 2.6 (p. 35) team questions for teachers to answer as part of the exercise.*	How will the team score assessment tasks? What is proficiency with a concept? How calibrated is team scoring? • Determine how to score a task. • Answer questions about scoring tasks on assessments. • Consider examples of scoring guides for tasks using points and a proficiency scale. • Use student work to calibrate scoring.	Team discussion tools: Figure 2.5: Teacher team discussion tool—Sample assessment questions for team scoring (page 34) Figure 2.6: Teacher team discussion tool—Scoring assessment tasks (page 35) Figure 2.7: Example of task scoring guides or proficiency rubric scores for tasks in figure 2.5 (pages 37–38) Figure 2.8: Sample assessment questions with student work for team scoring (pages 39–40)
Chapter 3: Sample Common Mathematics Assessments and Calibration Routines			
Explore examples of common assessments—both mid-unit and end-of-unit.	What do quality common mid-unit assessments look like? What do quality common end-of-unit assessments look like?	Figure 3.1: Grade 1 common mid-unit assessment place value sample (page 48) Figure 3.2: Grade 5 common mid-unit geometry assessment sample (page 49) Figure 3.3: Grade 8 common mid-unit assessment solving linear equations sample (page 50) Figure 3.4: Kindergarten counting and written numbers common assessment sample (pages 52–53) Figure 3.5: Sample grade 4 end-of-unit assessment (pages 54–57) Figure 3.6: Sample algebra 1 / integrated mathematics I end-of-unit assessment (pages 59–64)
Explore examples of scoring guides as team scoring agreements for common assessments.	How might the team document scoring guides for assessments?	Figure 3.7: Sample grade 4 fractions end-of-unit assessment scoring guide (pages 66–67) Figure 3.8: Sample algebra 1 / integrated mathematics I end-of-unit assessment scoring guide (pages 68–69)
Explore examples of proficiency scales as team scoring agreements for common assessments.	How should the team think about student learning for levels 1–4 on a proficiency scale? How can a proficiency scale be used for a group of questions related to the same standard?	Figure 3.9: Proficiency rubric for grade 1 common mid-unit assessment sample in figure 3.1 (page 70) Figure 3.10: Example proficiency rubrics for the grade 4 end-of-unit assessment in figure 3.5 (pages 71–72) Figure 3.11: Example proficiency rubrics for the algebra 1 / integrated mathematics I end-of-unit assessment in figure 3.6 (pages 72–73)

Determine how to commonly score assessments.	What must a student show to be proficient or earn points on a common assessment? How might partial credit be given consistently? Should questions on the assessment be revised?	Figure 3.12: Teacher team discussion tool—Assessment instrument alignment and scoring agreements (page 75)
Calibrate scoring of common assessments across the team.	How will everyone on the team score common assessments? What needs to be calibrated?	Figure 3.13: Teacher team discussion tool—Equitable scoring and calibration routine (page 78)
Chapter 4: Teacher Actions in the Formative Assessment Process		
Analyze specific learning trends at various levels of proficiency and create plans for re-engagement.	Do all team members have agreement on using student work at the levels of below, approaching, meeting, and exceeding proficiency? How should the team use the protocol?	Figure 4.1: Student Work Protocol (page 84) Figure 4.2: Student Work Protocol instructions (page 85)
Analyze proficiency levels of students by each essential standard and plan team next steps with instructions.	Do all team members know how many students are proficient with each essential learning standard? Do the teachers know the instructional strategies that are supporting proficient or not-proficient work? What are intervention or instruction plans for the next unit? How will the team use the protocol?	Figure 4.3: Essential Learning Standard Analysis Protocol—Grade 4 sample (pages 87–88) Figure 4.4: Essential Learning Standard Analysis Protocol instructions (page 89)
Analyze the complexity of thinking or the solution pathways that students use to problem solve with instructions.	Do all team members know the different problem-solving strategies students use? How will the team use the protocol?	Figure 4.5: Student Reasoning Protocol—Grade 2 example (pages 91–92) Figure 4.6: Student Reasoning Protocol instructions (page 93)
Chapter 5: Student Actions in the Formative Assessment Process		
Explore examples of student self-assessments after standards are learned.	How might the team create structures for student self-reflection by essential learning standard?	Figure 5.1: Sample student end-of-unit self-assessment—Grade 4 fractions unit (page 100) Figure 5.2: Sample student end-of-unit self-assessment—High school linear functions unit (page 100) Figure 5.3: Example kindergarten self-assessment (page 102) Figure 5.4: Student progress-tracking chart from algebra 1 in Norwalk Community School District (pages 103–104) Figure 5.5: Example grade 8 student contract and test correction form (pages 106–107)

Explore examples of student self-assessments during a unit using essential learning standards.	How might the team create structures for student self-reflection by essential learning standard throughout a unit?	Figure 5.6: Grade 7 student tracker—Self-assessment and action plan (pages 109–110)
		Figure 5.7: Grade 3 student tracker—Self-assessment and action plan (page 112)
		Figure 5.8: Grade 5 student tracker—Self-assessment and action plan (page 113)
		Figure 5.9: Grade 1 student tracker—Self-assessment and action plan (page 114)
		Figure 5.10: High school student tracker—Self-assessment and action plan (pages 115–116)
Determine team agreements regarding feedback to students and student actions.	How will the team calibrate scoring and give consistent feedback? How will students use the feedback for learning? How will students self-reflect?	Figure 5.11: Teacher team discussion tool—Collaborative team feedback with student action (page 118)
Chapter 6: Team Response to Student Learning Using Tier 2 Mathematics Intervention Criteria		
Diagram showing the formative assessment process in units.	When should the team analyze and take action on common assessment data? When should students take action on common assessment data?	Figure 6.1: Common assessment formative process for teachers and students (page 123)
Have the team answer questions about intervention practices.	Why do there need to be team interventions? How will the team plan for interventions? How will the team know if interventions are effective?	Figure 6.2: Teacher team discussion tool—Intervention practices (page 124)
Analyze and evaluate the team mathematics interventions.	Using the rubric, what is the "grade" for team intervention practices? What is a strength? What needs to be addressed?	Figure 6.3: Teacher team rubric—Tier 2 mathematics intervention program (page 125)
Have the team plan for a team intervention.	What are the interventions needed, and which students are in each? What will students learn?	Figure 6.4: Tier 1 and Tier 2 schedule options (page 126) Figure 6.5: Teacher team discussion tool—Mathematics intervention planning tool (page 129)

Use the QR code to access a downloadable version of this appendix and additional online resources and examples.

References and Resources

Adlai E. Stevenson High School. (2022, August). *Course description: MTH 151 / MTH 152—Algebra 1*. Accessed at https://docs.google.com/document/d/1MTpuAQeM2rNvQjc5v4kiNVZlADqrAdMMV4PS4aPprOk/edit on November 14, 2022.

Aguirre, J., Mayfield-Ingram, K., & Martin, D. B. (2013). *The impact of identity in K–8 mathematics: Rethinking equity-based practices*. Reston, VA: National Council of Teachers of Mathematics.

Ainsworth, L. (2003). *"Unwrapping" the standards: A simple process to make standards manageable*. Englewood, CO: Advanced Learning Press.

AllThingsPLC. (n.d.). *Evidence of effectiveness: Field Elementary*. Accessed at www.allthingsplc.info/evidence/details/id,998 on September 19, 2017.

Bailey, K., & Jakicic, C. (2012). *Common formative assessment: A toolkit for Professional Learning Communities at Work*. Bloomington, IN: Solution Tree Press.

Bailey, K., & Jakicic, C. (2019). *Make it happen: Coaching with the four critical questions of PLCs at Work*. Bloomington, IN: Solution Tree Press.

Bailey, K., & Jakicic, C. (2023). *Common formative assessment: A toolkit for Professional Learning Communities at Work* (2nd ed.). Bloomington, IN: Solution Tree Press.

Barth, R. S. (2001). *Learning by heart*. San Francisco: Jossey-Bass.

Black, P., & Wiliam, D. (2001). *Inside the black box: Raising standards through classroom assessment*. London: Assessment Group of the British Educational Research Association.

Black, P., & Wiliam, D. (2010). Inside the black box: Raising standards through classroom assessment. *Phi Delta Kappan, 92*(1), 81–90.

Boston, M., Dillon, F., Smith, M. S., & Miller, S. (2017). *Taking action: Implementing effective mathematics teaching practices, grades 9–12*. Reston, VA: National Council of Teachers of Mathematics.

Brown, T., & Ferriter, W. M. (2021). *You can learn! Building student ownership, motivation, and efficacy with the PLC at Work process*. Bloomington, IN: Solution Tree Press.

Buffum, A., Mattos, M., & Malone, J. (2018). *Taking action: A handbook for RTI at Work*. Bloomington, IN: Solution Tree Press.

Chappuis, J., & Stiggins, R. (2020). *Classroom assessment for student learning: Doing it right—using it well* (3rd ed.). New York: Pearson.

Chappuis, S., & Stiggins, R. (2002). Classroom assessment for learning. *Educational Leadership, 60*(1), 40–43.

Covey, S. R. (1989). *The seven habits of highly effective people: Restoring the character ethic*. New York: Simon & Schuster.

Covey, S. R. (2004). *The seven habits of highly effective people: Restoring the character ethic* (Rev. ed.). New York: Free Press.

DuFour, R. (2015). *In praise of American educators: And how they can become even better*. Bloomington, IN: Solution Tree Press.

DuFour, R., DuFour, R., Eaker, R., & Karhanek, G. (2009). *Raising the bar and closing the gap: Whatever it takes*. Bloomington, IN: Solution Tree Press.

DuFour, R., DuFour, R., Eaker, R., Many, T. W., & Mattos, M. (2016). *Learning by doing: A handbook for Professional Learning Communities at Work* (3rd ed.). Bloomington, IN: Solution Tree Press.

DuFour, R., & Marzano, R. J. (2011). *Leaders of learning: How district, school, and classroom leaders improve student achievement.* Bloomington, IN: Solution Tree Press.

Eaker, R., & Keating, J. (2012). *Every school, every team, every classroom: District leadership for growing Professional Learning Communities at Work.* Bloomington, IN: Solution Tree Press.

Erkens, C., Schimmer, T., & Dimich, N. (2017). *Essential assessment: Six tenets for bringing hope, efficacy, and achievement to the classroom.* Bloomington, IN: Solution Tree Press.

Fernandes, A., Crespo, S., & Civil, M. (2017). Introduction. In A. Fernandes, S. Crespo, & M. Civil (Eds.), *Access and equity: Promoting high-quality mathematics, grades 6–8* (pp. 1–10). Reston, VA: National Council of Teachers of Mathematics.

Gallimore, R., Ermeling, B. A., Saunders, W. M., & Goldenberg, C. (2009). Moving the learning of teaching closer to practice: Teacher education implications of school-based inquiry teams. *Elementary School Journal, 109*(5), 537–553.

Guarino, J., Cole, S., & Sperling, M. (2022). Our children are not numbers. *Mathematics Teacher: Learning and Teaching PK–12, 115*(6), 404–412.

Guskey, T. R. (Ed.). (2009). *The teacher as assessment leader.* Bloomington, IN: Solution Tree Press.

Hansen, A. (2015). *How to develop PLCs for singletons and small schools.* Bloomington, IN: Solution Tree Press.

Hattie, J. A. C. (2009). *Visible learning: A synthesis of over 800 meta-analyses relating to achievement.* New York: Routledge.

Hattie, J. A. C. (2012). *Visible learning for teachers: Maximizing impact on learning.* New York: Routledge.

Hattie, J. A. C., Fisher, D., & Frey, N. (2017). *Visible learning for mathematics, grades K–12: What works best to optimize student learning.* Thousand Oaks, CA: Corwin Mathematics.

Jilk, L. M., & Erickson, S. (2017). Shifting students' beliefs about competence by integrating mathematics strengths into tasks and participation norms. In A. Fernandes, S. Crespo, & M. Civil (Eds.), *Access and equity: Promoting high-quality mathematics, grades 6–8* (pp. 11–26). Reston, VA: National Council of Teachers of Mathematics.

Kanold, T. D., Barnes, B., Larson, M. R., Kanold-McIntyre, J., Schuhl, S., & Toncheff, M. (2018). *Mathematics homework and grading in a PLC at Work.* Bloomington, IN: Solution Tree Press.

Kanold, T. D. (Ed.), Briars, D. J., Asturias, H., Foster, D., & Gale, M. A. (2013). *Common Core mathematics in a PLC at Work, grades 6–8.* Bloomington, IN: Solution Tree Press.

Kanold, T. D. (Ed.), Dixon, J. K., Adams, T. L., & Nolan, E. C. (2015). *Beyond the Common Core: A handbook for mathematics in a PLC at Work, grades K–5.* Bloomington, IN: Solution Tree Press.

Kanold, T. D., Kanold-McIntyre, J., Larson, M. R., Barnes, B., Schuhl, S., & Toncheff, M. (2018). *Mathematics instruction and tasks in a PLC at Work.* Bloomington, IN: Solution Tree Press.

Kanold, T. D. (Ed.), Kanold-McIntyre, J., Larson, M. R., & Briars, D. J. (2015). *Beyond the Common Core: A handbook for mathematics in a PLC at Work, grades 6–8.* Bloomington, IN: Solution Tree Press.

Kanold, T. D. (Ed.), & Larson, M. R. (2012). *Common Core mathematics in a PLC at Work, leader's guide.* Bloomington, IN: Solution Tree Press.

Kanold, T. D. (Ed.), & Larson, M. R. (2015). *Beyond the Common Core: A handbook for mathematics in a PLC at Work, leader's guide.* Bloomington, IN: Solution Tree Press.

Kanold, T. D. (Ed.), Larson, M. R., Fennell, F., Adams, T. L., Dixon, J. K., Kobett, B. M., & Wray, J. A. (2012a). *Common Core mathematics in a PLC at Work, grades K–2.* Bloomington, IN: Solution Tree Press.

Kanold, T. D. (Ed.), Larson, M. R., Fennell, F., Adams, T. L., Dixon, J. K., Kobett, B. M., & Wray, J. A. (2012b). *Common Core mathematics in a PLC at Work, grades 3–5.* Bloomington, IN: Solution Tree Press.

Kanold, T. D. (Ed.), & Toncheff, M. (2015). *Beyond the Common Core: A handbook for mathematics in a PLC at Work, high school.* Bloomington, IN: Solution Tree Press.

Kanold, T. D., Toncheff, M., Larson, M. R., Barnes, B., Kanold-McIntyre, J., & Schuhl, S. (2018). *Mathematics coaching and collaboration in a PLC at Work.* Bloomington, IN: Solution Tree Press.

Kanold, T. D. (Ed.), Zimmermann, G., Carter, J. A., Kanold, T. D., & Toncheff, M. (2012). *Common Core mathematics in a PLC at Work, high school*. Bloomington, IN: Solution Tree Press.

Kramer, S. V. (2015). *How to leverage PLCs for school improvement*. Bloomington, IN: Solution Tree Press.

Kramer, S. V., & Schuhl, S. (2017). *School improvement for all: A how-to-guide for doing the right work*. Bloomington, IN: Solution Tree Press.

Leane, B., & Yost, J. (2022). *Singletons in a PLC at Work: Navigating on-ramps to meaningful collaboration*. Bloomington, IN: Solution Tree Press.

Mattos, M. (2016, November 2). *Connecting PLCs and RTI* [Blog post]. Accessed at www.allthingsplc.info/blog/view/335/connecting-plcs-and-rti on March 24, 2023.

Mattos, M., DuFour, R., DuFour, R., Eaker, R., & Many, T. W. (2016). *Concise answers to frequently asked questions about Professional Learning Communities at Work*. Bloomington, IN: Solution Tree Press.

Missoula County Public Schools Curriculum Consortium. (2013, June). *PreK–12 mathematics curriculum*. Missoula, MT: Missoula County Public Schools. Accessed at www.mcpsmt.org/cms/lib03/MT01001940/Centricity/Domain/825/Mathematics%20Curriculum.pdf on March 5, 2023.

Moss, C. M., & Brookhart, S. M. (2019). *Advancing formative assessment in every classroom: A guide for instructional leaders* (2nd ed.). Alexandria, VA: ASCD.

National Council of Supervisors of Mathematics. (n.d.). *Illustrating the Standards for Mathematical Practice: Module index*. Accessed at www.mathedleadership.org/ismp on March 5, 2023.

National Council of Supervisors of Mathematics. (2015). *Improving student achievement by infusing highly effective instructional strategies into multi-tiered support systems (MTSS): Response to intervention (RTI) Tier 2 instruction*. Accessed at www.mathedleadership.org/member/docs/resources/positionpapers/NCSMPositionPaper15.pdf on June 1, 2017.

National Council of Supervisors of Mathematics. (2022). *Culturally relevant leadership in mathematics*. Aurora, CO: Author.

National Council of Teachers of Mathematics. (2014). *Principles to actions: Ensuring mathematical success for all*. Reston, VA: Author.

National Council of Teachers of Mathematics. (2018). *Catalyzing change in high school mathematics: Initiating critical conversations*. Reston, VA: Author.

National Council of Teachers of Mathematics. (2020). *Catalyzing change in middle school mathematics: Initiating critical conversations*. Reston, VA: Author.

National Governors Association Center for Best Practices & Council of Chief State School Officers. (2010). *Common Core State Standards for mathematics*. Washington, DC: Authors. Accessed at https://ccsso.org/sites/default/files/2017-10/MathStandards50805232017.pdf on March 5, 2023.

National Science Digital Library. (n.d.). *Search results: Mathematics*. Accessed at https://nsdl.oercommons.org/browse?f.general_subject=mathematics on September 19, 2017.

Popham, W. J. (2011). *Transformative assessment in action: An inside look at applying the process*. Alexandria, VA: ASCD.

Reeves, D. (2002). *The leader's guide to standards: A blueprint for educational equity and excellence*. San Francisco: Jossey-Bass.

Reeves, D. (2011). *Elements of grading: A guide to effective practice*. Bloomington, IN: Solution Tree Press.

Reeves, D. (2016). *Elements of grading: A guide to effective practice* (2nd ed.). Bloomington, IN: Solution Tree Press.

Safir, S., & Dugan, J. (2021). *Street data: A next-generation model for equity, pedagogy, and school transformation*. Thousand Oaks, CA: Corwin.

Schimmer, T. (2016). *Grading from the inside out: Bringing accuracy to student assessment through a standards-based mindset*. Bloomington, IN: Solution Tree Press.

Schuhl, S., Kanold, T. D., Barnes, B., Jain, D. M., Larson, M. R., & Mozingo, B. (2021). *Mathematics unit planning in a PLC at Work, high school*. Bloomington, IN: Solution Tree Press.

Schuhl, S., Kanold, T. D., Deinhart, J., Lang-Raad, N. D., Larson, M. R., & Smith, N. N. (2021). *Mathematics unit planning in a PLC at Work, grades preK–2*. Bloomington, IN: Solution Tree Press.

Schuhl, S., Kanold, T. D., Deinhart, J., Larson, M. R., & Toncheff, M. (2020). *Mathematics unit planning in a PLC at Work, grades 3–5*. Bloomington, IN: Solution Tree Press.

Schuhl, S., Kanold, T. D., Kanold-McIntyre, J., Chuang, S., Larson, M. R., & Smith, M. (2021). *Mathematics unit planning in a PLC at Work, grades 6–8*. Bloomington, IN: Solution Tree Press.

Smarter Balanced. (n.d.). *Sample index*. Accessed at http://sampleitems.smarterbalanced.org/BrowseItems on September 19, 2017.

Smith, M. S., Steele, M., & Raith, M. L. (2017). *Taking action: Implementing effective mathematics teaching practices, grades 6–8*. Reston, VA: National Council of Teachers of Mathematics.

Smith, M. S., & Stein, M. K. (1998). Selecting and creating mathematical tasks: From research to practice. *Mathematics Teaching in the Middle School, 3*(5), 344–350.

Smith, M. S., & Stein, M. K. (2011). *Five practices for orchestrating productive mathematics discussions*. Reston, VA: National Council of Teachers of Mathematics.

Smith, M. S., & Stein, M. K. (2018). *Five practices for orchestrating productive mathematics discussions* (2nd ed.). Reston, VA: National Council of Teachers of Mathematics.

Stiggins, R. (2007). Assessment through the student's eyes. *Educational Leadership, 64*(8), 22–26. Accessed at www.ascd.org/el/articles/assessment-through-the-students-eyes-summer-2007 on March 5, 2023.

Stiggins, R., Arter, J. A., Chappuis, J., & Chappuis, S. (2006). *Classroom assessment for student learning: Doing it right—using it well*. Princeton, NJ: Educational Testing Service.

Toncheff, M., Kanold, T. D., Schuhl, S., Barnes, B., Deinhart, J., Kanold-McIntyre, J., & Larson, M. R. (in press). *Mathematics instruction and tasks in a PLC at Work* (2nd ed.). Bloomington, IN: Solution Tree Press.

Wiliam, D. (2011). *Embedded formative assessment*. Bloomington, IN: Solution Tree Press.

Wiliam, D. (2016). The secret of effective feedback. *Educational Leadership, 73*(7), 10–15.

Wiliam, D. (2018). *Embedded formative assessment* (2nd ed.). Bloomington, IN: Solution Tree Press.

Index

A
academic language, 40, 41–42, 44
accommodations and modifications, 43, 128
assessment instrument quality evaluation rubric
    criteria for, 21, 40
    designing common assessments and, 44
    end-of-unit assessments and, 76
    essential learning standards and, 21
    mid-unit assessments and, 53
    teacher team rubric for, 18
assessment instruments
    about this book, 4
    assessment process versus, 7–9
    equity and, 6
    feedback and, 10
    formative assessment process and, 135
    manipulatives and technology and, 43
    quality common mathematics assessments and, 17
    reflecting on team assessment practices and, 13
    sample common mathematics assessments and, 65
    teacher team discussion tool—assessment instrument alignment and scoring agreements, 75
assessments. *See also* common assessments; end-of-unit assessments; mid-unit assessments; self-assessments
    assessment instrument versus process, 7–9
    formats of assessment tasks, 30–32
    impact of, xi
    one-on-one assessments, 42, 78
    reflection on team assessment practices, 13–15
    rolling assessments, 51, 52–53, 111
    teacher team discussion tool—high-quality assessment evaluation, 19–20
    team recommendations for making a plan for assessment methods, 32

B
Bailey, K., 9
Barnes, B., 98
Barth, R., 81
big ideas of a PLC, 1

C
calibration. *See also* sample common mathematics assessments and calibration routines
    quality of assessment feedback and, 76, 77
    teacher team discussion tool—equitable scoring and calibration routines, 78
    team recommendations for using calibration routines for common unit assessments, 80
Catalyzing Change in Middle School Mathematics (NCTM), 82
Chappuis, J., 9
clarity of directions, 41, 44
cognitive demand
    balance of higher- and lower-level-cognitive-demand mathematical tasks, 26, 29–30
    cognitive-demand-level task analysis guide, 137
    examples of lower- and higher-level-cognitive demand tasks, 27–28
    team recommendations for agreeing on lower- and higher-level-cognitive-demand mathematical tasks chosen for common unit assessments, 30
Cole, S., 82
common assessments
    about, 5
    clarity of directions, academic language, visual presentation, and logistics and, 41–45
    common assessment formative process, 12, 123, 135
    criteria to consider when designing, 44–45
    four specific purposes of, 5–7
    Mathematics at Work common assessment process. *See* Mathematics at Work common assessment process

PLCs at Work and, 82
quality common mathematics assessments. *See* quality common mathematics assessments
sample common mathematics assessments. *See* sample common mathematics assessments and calibration routines
teacher team discussion tool—formative use of common assessments, 10
team action 1 and, 4, 13, 45, 135
team recommendations for designing high-quality common unit assessments, 21
team recommendations for knowing the purpose of common assessments, 7
common formative assessments (CFAs), 8, 111, 117
conceptual understanding, 29, 43, 132, 133
critical questions of a PLC. *See also* PLCs (Professional Learning Communities) at Work
about, 1–2
common assessments and, 6
essential learning standards and, 22
evidence of student learning protocols and, 82, 83
reflection on team assessment practices and, 13, 15
scoring agreements and, 75
self-regulatory feedback and, 117
student action after common mid-unit assessments and, 108

D

data analysis
about, 81–82
evidence of student learning protocols/data dialogue protocols. *See* evidence of student learning protocols
formative assessment process and, 135
team action 2 and, 4, 15, 80, 135
team recommendations for engaging in teacher team data-analysis and action routines, 95
data/map data, 95
directions, clarity of, 41, 44
double scoring, 77. *See also* scoring agreements
DuFour, R., 9

E

end-of-unit assessments. *See also* assessments
example of, 54–57, 58–64
example proficiency rubrics for, 71–72, 72–73
example scoring guides for, 66–67, 68–69
example student end-of-unit self-assessment, 100
logistics of quality common assessments and, 42
sample common end-of-unit assessments, 53, 65
student action after, 97–99, 101–102, 105, 108
team recommendations for creating common end-of-unit assessments, 76
team recommendations for requiring student action on feedback from common end-of-unit assessments, 105
equity
common assessments and, 6, 21
PLCs and, 1–2

essential learning standard analysis protocol. *See also* evidence of student learning protocols
example of, 87–88, 89
use of, 85–86, 89–90
essential learning standards
example of, 24–25
identification of and emphasis on, 21–23, 26
student trackers and, 108
team recommendations for determining essential learning standards for student assessment and reflection, 26
Tier 2 interventions and, 128, 130
evidence of student learning and Tier 2 interventions, 132–134
evidence of student learning protocols. *See also* data analysis
about, 82–83
essential learning standard analysis protocol, 85–90
student reasoning protocol, 90–95
student work protocol, 83–85

F

feedback
assessment instrument versus process and, 9
calibration routines and, 76–79
evaluation of mathematics assessment feedback processes, 10–11
FAST feedback, 9
formative assessment process and, 135
just-in-time feedback, 131
quality student feedback, 79
sample common mathematics assessments and calibration routines and, 76–80
scoring and, 36
self-regulatory feedback, 117, 119
student actions and, 99, 102
student-to-student feedback, 105
team recommendations for creating a system for student action and ownership of learning based on formative feedback during a unit using common mid-unit assessments, 119
team recommendations for requiring student action on feedback from common end-of-unit assessments, 105
Fisher, D., 79
formative assessment process
common assessment formative process, 12, 123, 135
student actions in the formative assessment process. *See* student actions in the formative assessment process
teacher actions in the formative assessment process. *See* data analysis
formats of assessment tasks, 30–32
Frey, N., 79

G

goals
of assessment process, 108
collaborative culture and, 1
essential learning standards and, 22
feedback and, 11, 76
formative assessment process and, 98–99, 101, 102, 135

guaranteed and viable curriculum, 22
Guarino, J., 82

H
Hattie, J., 79
higher-level-cognitive-demand mathematical tasks. *See also* cognitive demand
 balance of lower-level-cognitive-demand tasks and, 26, 29–30
 cognitive-demand-level task analysis guide, 137
 examples of, 27–28
 logistics of quality common assessments and, 42
 student reasoning protocol and, 90
 team recommendations for agreeing on lower- and higher-level-cognitive-demand mathematical tasks chosen for common unit assessments, 30

I
I can statements, 23
inter-rater reliability, 77
interventions
 about, 121–122, 124
 common assessments and, 7
 criteria to consider for, 124
 essential learning standards and, 128, 130
 fluid and flexible, 130
 formative assessment process and, 135
 just in time, 131–132
 proven to show evidence of student learning, 132–134
 systematic and required, 126–128
 teacher team discussion tool—intervention practices, 124
 teacher team discussion tool—mathematics intervention planning tool, 129
 teacher team rubric—tier 2 mathematics intervention program, 125
 team action 2 and, 4, 15, 80, 135
 team recommendations for creating a Tier 2 intervention program, 134
 tier 1 and tier 2 schedule options, 126
introduction
 about equity and mathematics assessment and intervention, 1
 about this book, 4
 equity and PLCs, 1–2
 mathematics in a PLC at Work framework, 2–3
 reflect, refine, and act cycle, 2

J
Jakicic, C., 9

K
Kanold, T., 17, 127
Kanold-McIntyre, J., 11

L
learning targets
 essential learning standards and, 23
 student trackers and, 108

logistics of quality common assessments, 42–43, 44
lower-level-cognitive-demand mathematical tasks. *See also* cognitive demand
 balance of higher-level-cognitive-demand tasks and, 26, 29–30
 cognitive-demand-level task analysis guide, 137
 examples of lower- and higher-level-cognitive demand tasks, 27–28
 team recommendations for agreeing on lower- and higher-level-cognitive-demand mathematical tasks chosen for common unit assessments, 30

M
manipulatives (logistics of quality common assessments), 43
map data, 95
matching assessment tasks, 31. *See also* assessments
mathematics assessment and intervention in a PLC at Work protocols and tools reproducible, 138–141
Mathematics at Work common assessment process. *See also* common assessments; sample common mathematics assessments and calibration routines
 about, 5
 assessment instrument versus the assessment process, 7–9
 evaluation of mathematics assessment feedback processes, 10–11
 FAST feedback, 9
 four specific purposes of, 5–7
 reflection on team assessment practices, 13–15
mid-unit assessments. *See also* assessments
 example of, 48, 49, 50
 example proficiency rubrics for, 70
 sample common mid-unit assessments, 47, 51, 53
 student action after, 108
 team recommendations for creating common mid-unit assessments, 53
multiple choice assessment tasks, 31. *See also* assessments

N
National Council of Teachers of Mathematics (NCTM), 82, 108, 122, 133

O
one-on-one assessments, 42, 78. *See also* assessments
"Our Children Are Not Numbers" (Guarino, Cole, and Sperling), 82

P
PLCs (Professional Learning Communities) at Work
 collaboration and, xii
 common assessments and, 82
 critical questions of a PLC. *See* critical questions of a PLC
 equity and PLCs, 1–2
 mathematics assessment and intervention in a PLC at Work protocols and tools reproducible, 138–141
 mathematics in a PLC at Work framework, 2–3
procedural fluency, 29, 128, 133
professional development, four specific purposes of common assessments, 6–7

proficiency maps, 22
proficiency rubrics
    example of, 70, 71–72, 72–73
    example of tasks scoring guides or proficiency rubric scores, 37–38
    scoring and, 33, 70, 74–76

Q

quality common mathematics assessments. *See also* common assessments
    about, 17, 21
    assessment-task formats and, 30–32
    clarity of directions, academic language, visual presentation, and logistics, 41–45
    cognitive-demand tasks, balance of higher- and lower-level-, 26–30
    essential learning standards and, 21–26
    scoring agreements and, 32–40
    teacher team discussion tool—high-quality assessment evaluation, 19–20
    teacher team rubric—assessment instrument quality evaluation, 18
    team recommendations for designing high-quality common unit assessments, 21

R

Raith, M., 22
reflect, refine, and act cycle, 2
reflection
    essential learning standards and, 23, 128
    formative assessment process and, 134, 135
    goal setting and, 99
    reflection on team assessment practices, 13
    student actions and, 97, 98, 101–103, 117
    student ownership of learning and, 7
    teacher team discussion tool—common assessment self-reflection protocol, 14
reproducible for mathematics assessment and intervention in a PLC at Work protocols and tools, 138–141
rigor-balance ratio, 29. *See also* cognitive demand
Rivera, G., 94
rolling assessments, 51, 52–53, 111. *See also* assessments
RTI (response to intervention), 121

S

sample common mathematics assessments and calibration routines. *See also* common assessments; Mathematics at Work common assessment process
    about, 47
    collaborative scoring agreements, 66–76
    feedback and, 76–80
    sample common end-of-unit assessments, 53–65
    sample common mid-unit assessments, 47–53
    teacher team discussion tool—assessment instrument alignment and scoring agreements, 75
    teacher team discussion tool—equitable scoring and calibration routines, 78
Schuhl, S., personal stories of, 6, 8, 74, 77, 86, 101, 131
scoring agreements
    collaborative scoring agreements, 66–76
    example of tasks scoring guides or proficiency rubric scores, 37–38
    proficiency rubrics and, 70–76
    quality common mathematics assessments and, 32–33, 36, 40
    sample assessment questions with student work for team scoring, 39–40
    scoring guides and, 66–69
    teacher team discussion tool—assessment instrument alignment and scoring agreements, 75
    teacher team discussion tool—equitable scoring and calibration routines, 78
    teacher team discussion tool—sample assessment questions for scoring, 34
    teacher team discussion tool—scoring assessment tasks, 35
    team recommendations for determining common scoring agreements for common mathematics assessments, 40
scoring guides
    about, 66
    example of, 66–67, 68–69
self-assessments. *See also* assessments
    example of, 100, 102
    example of student trackers with self-assessment and action plan, 109–110, 112, 113, 114, 115–116
    goals of assessment and, 108
    student actions and, 97, 99, 101
    student trackers and, 111
self-reflection. *See* reflection
self-regulatory feedback, 117, 119
singleton teachers
    calibration routines and, 79
    cognitive-demand mathematical tasks and, 30
    end-of-unit assessments and, 65
    mathematics in a PLC at Work framework, 3
    scoring guides and, 66
Smith, M., 22
Sperling, M., 82
standards, 22. *See also* essential learning standards
Steele, M., 22
Stiggins, R., 9
student actions in the formative assessment process. *See also* end-of-unit assessments; mid-unit assessments
    about, 97
    after common end-of-unit assessments, 97–102, 105, 108
    after common mid-unit assessments, 108
    example student contract and test correction form, 106–107
    example student progress-tracking chart, 103–104
    self-regulatory feedback, 117, 119
    student trackers, 108–117, 115–116
    teacher team discussion tool—collaborative team feedback with student action, 118

team recommendations for requiring student action on feedback from common end-of-unit assessments, 105
student ownership of learning
    feedback and, 10
    four specific purposes of common assessments, 5, 7
    student actions in the formative assessment process and, 99, 101, 108
    team recommendations for, 119
student reasoning protocol. *See also* evidence of student learning protocols
    example of, 91–92
    student reasoning protocol instructions, 93
    use of, 90, 93
student trackers
    example of with self-assessment and action plan, 109–110, 112, 113, 115–116
    example student progress-tracking chart, 103–104
    student actions in the formative assessment process and, 108, 111, 114–117
student work protocol, 83–85. *See also* evidence of student learning protocols
summative grades, 99

T

teacher actions in the formative assessment process. *See* data analysis
teacher teams
    creating teams for singletons, 3
    mathematics in a PLC at Work framework and, 3
    teacher team discussion tool—assessment instrument alignment and scoring agreements, 75
    teacher team discussion tool—collaborative team feedback with student action, 118
    teacher team discussion tool—common assessment self-reflection protocol, 14
    teacher team discussion tool—equitable scoring and calibration routines, 78
    teacher team discussion tool—formative use of common assessments, 10
    teacher team discussion tool—high-quality assessment evaluation, 19–20
    teacher team discussion tool—intervention practices, 124
    teacher team discussion tool—mathematics intervention planning tool, 129
    teacher team discussion tool—sample assessment questions for scoring, 34
    teacher team discussion tool—scoring assessment tasks, 35
    teacher team rubric—assessment instrument quality evaluation, 18
    teacher team rubric—common assessment formative process, 12
    teacher team rubric—tier 2 mathematics intervention program, 125
    team action 1 and, 4, 13, 45, 135
    team action 2 and, 4, 15, 80, 135
team recommendations for
    agreeing on lower- and higher-level-cognitive-demand mathematical tasks chosen for common unit assessments, 30
    creating common end-of-unit assessments, 76
    creating common mid-unit assessments, 53
    creating a system for student action and ownership of learning based on formative feedback during a unit using common mid-unit assessments, 119
    creating a Tier 2 mathematics intervention program, 134
    designing high-quality common unit assessments, 21
    determining clarity of directions, academic language, visual presentation, and logistics, 44
    determining common scoring agreements for common mathematics assessments, 40
    determining essential learning standards for student assessment and reflection, 26
    engaging in teacher team data-analysis and action routines, 95
    knowing the purpose of common assessments, 7
    making a plan for assessment methods, 32
    requiring student action on feedback from the common end-of-unit assessment, 105
    using calibration routines for common unit assessments, 80
team response to student learning using Tier 2 mathematics intervention criteria. *See* interventions
technology (logistics of quality common assessments), 43
Toncheff, M., 83, 123

V

visual representations, 42, 44

W

Wiliam, D., 99
WIN process, 122, 123

Y

Yost, J., 81

Mathematics Instruction and Tasks in a PLC at Work®, Second Edition
Mona Toncheff, Timothy D. Kanold, Sarah Schuhl, Bill Barnes, Jennifer Deinhart, Jessica Kanold-McIntyre, and Matthew R. Larson
Build collective teacher efficacy and students' mathematical thinking using the Mathematics in a PLC at Work™ lesson-design process. PreK–12 teacher teams will learn new and enhanced research-affirmed lesson-design elements and how to efficiently elicit high levels of student engagement and self-efficacy.
BKG147

Mathematics Unit Planning in a PLC at Work®, Grades PreK–2
Sarah Schuhl, Timothy D. Kanold, Jennifer Deinhart, Nathan D. Lang-Raad, Matthew R. Larson, and Nanci N. Smith
Discover how to fully answer PLC critical question one in your mathematics classroom: what do we want all students to know and be able to do? With this resource, your teacher team will acquire detailed model mathematics units, learn how to perform seven collaborative tasks, and more.
BKF964

Mathematics Unit Planning in a PLC at Work®, Grades 3–5
Sarah Schuhl, Timothy D. Kanold, Jennifer Deinhart, Matthew R. Larson, and Mona Toncheff
Part of the *Every Student Can Learn Mathematics* series, this practical resource provides a framework for collectively planning a unit of study in grades 3–5. Grade-level teams learn to work together to unwrap standards, create team unit calendars, design robust fraction units, and more.
BKF965

Mathematics Unit Planning in a PLC at Work®, Grades 6–8
Sarah Schuhl, Timothy D. Kanold, Jessica Kanold-McIntyre, Suyi Chuang, Matthew R. Larson, and Mignon Smith
Improve mathematics achievement across grades 6–8 through a collaborative unit planning process. With the authors' expert guidance, your team will learn how to collectively generate essential learning standards, create a unit calendar, identify prior knowledge, and complete many other vital collaborative tasks.
BKF966

Mathematics Unit Planning in a PLC at Work®, High School
Sarah Schuhl, Timothy D. Kanold, Bill Barnes, Darshan M. Jain, Matthew R. Larson, and Brittany Mozingo
Champion student mastery of essential mathematics content in grades 9–12. Part of the Every Student Can Learn Mathematics series, this guidebook provides high school teacher teams with a framework for collectively planning units of study.
BKF967

Solution Tree | Press *a division of Solution Tree*

Visit SolutionTree.com or call 800.733.6786 to order.

Wait! Your professional development journey doesn't have to end with the last pages of this book.

We realize improving student learning doesn't happen overnight. And your school or district shouldn't be left to puzzle out all the details of this process alone.

No matter where you are on the journey, we're committed to helping you get to the next stage.

Take advantage of everything from **custom workshops** to **keynote presentations** and **interactive web and video conferencing**. We can even help you develop an action plan tailored to fit your specific needs.

Let's get the conversation started.

Call 888.763.9045 today.

Solution Tree's mission is to advance the work of our authors. By working with the best researchers and educators worldwide, we strive to be the premier provider of innovative publishing, in-demand events, and inspired professional development designed to transform education to ensure that all students learn.

The National Council of Teachers of Mathematics advocates for high-quality mathematics teaching and learning for each and every student.